Scientific Work and Creativity: Advice from the Masters

A Compilation by the Citizen Scientists League

Edited by Reginald D. Smith

Cover Art: *A Philosopher Lecturing on the Orrery* by Joseph Wright of Derby (c. 1766), oil on canvas Derby Museum and Art Gallery. Public domain image courtesy of Wikimedia.

First Printing, 2012

ISBN: 978-0615644011

Citizen Scientists League
14881 Evergreen Avenue
Clearwater, FL 33762

http://citizenscientistsleague.com

Forward by Sheldon Greaves

Thomas Edison's famous remark that genius is 1 percent inspiration and 99 percent perspiration, in my opinion, gives creativity a bad rap. It tends to blind us to a crucial fact about scientific endeavor: inspiration, although insufficient, is necessary. Even the dogged slogging that defined Edison's search for the right combination of components for a practical incandescent light bulb demanded creativity on his part.

But in spite of creativity's essential role in scientific discovery, the nature of creativity resists easy explanation. We all know what it looks like; there is that subtle shock to the mind when we encounter an idea or invention that is truly creative. Some people who routinely exercise their creativity will describe a sense of suspended time and other pleasant sensations that accompany creative work, while others insist that it is dogged hard work. Still others feel driven to creative endeavor in spite of frustrations; they engage in their work because the frustration of not working exceeds that of working.

A larger question for those of us who aspire to a higher level of intellectual and artistic practice is, can creativity be taught? Can it be cultivated? Is there a strategy one can follow? Does one work for great mental breakthroughs, or generate a stream of ideas from which to sluice the nuggets? Is imagination, as Einstein declared, really more important than knowledge? Given what we know—or don't know—about the human mind, what can we say about the limits of human creativity?

Reginald Smith has done us a great service by bringing together a collection of some of the most productive minds of science writing about creativity. It is an area of study that receives considerable, but in my view insufficient attention. We still labor under the uncertainty at large that creativity is one of those things in which you either have it or you don't. I personally do not believe that this is true; I suspect it is a consequence of our imperfect understanding of how creativity works.

I am pleased to commend to the reader the essays in this book, if only because if one is to tease out the secrets of how the universe works, one must approach the level of creativity by which nature brought it about. This is not a trivial undertaking. It does, however, launch the persistent inquirer on the journey of a lifetime.

Sheldon Greaves, Ph.D., Director
Citizen Scientists League
May 22, 2012

Introduction from the Editor

This book is the culmination of about ten years of work gathering articles on scientific creativity by past and present great scientists. The editor first became fascinated with scientific creativity in college, where as a physics student, he read the biographies of many great scientists and books about scientific creativity such as *Discovering* by Robert S. Root-Bernstein. One of the most fascinating things was that while the scientific method was essential to doing actual scientific work, oftentimes insights sprung from a well of creativity that was vastly different than the prosaic accounts that many learn in high school or popular culture.

The great scientists themselves were open about how this sometimes abstract method of 'eureka', imagination, trial and error, and subconscious insight was crucial to the process of making the connections needed to discover new concepts and ideas. This is often different than the way we are taught how science is done as a methodical process from A to B. New discoveries are often made by indirect routes or accidentally when looking at a totally unrelated question.

This volume is meant for all those interested in the personal part of the scientific process and how they can use it to their advantage. There are essays by many great scientists, in fact some of the most renowned, which are an inside look at the style of genius which is sometimes written off as near mystical. Though there are common themes, some of them disagree or hold views we might feel are strange today. The insistence of Santiago Ramón y Cajal that the nature of matter is incapable of being solved by human minds (though he corrects this in the notes), the strident Marxism of J.B.S. Haldane, and the talk about a (then) much poorer and undeveloped India by C.V. Raman are obviously dated but are included uncensored in the interest of allowing the authors' full ideas to be seen.

Others disagree on methods of discovery. As an example compare the emphasis on intuition by Poincaré or the emphasis on chance by Mach and W.I.B. Beveridge with the insistences on method and hard work by Ostwald and Hamming. In the end though, all they say is illuminating.

This book is published by the Citizen Scientists League, an organization of both amateurs and professional scientists who are dedicated to helping all people do good scientific work, from participating in animal counting studies to publishing work in prominent peer-reviewed journals. It is our hope that this helps many aspiring, and even successful, scientific researchers from all walks of life.

We would like to thank the copyright owners for graciously granting us permission to reprint their works. Especially Harumi Yukawa, son of the physicist Hideki Yukawa, with whom we communicated across language barriers to receive permission. Thanks in this aspect also belongs to Michiji Konuma, head of the World Peace Group, who acted as an intermediary. Also, we would like to thank the wonderful work of Google (Google Books) and the Biodiversity Heritage Library whose online books allowed us to procure much of the public domain material with ease. Finally, we would like to thank Quintus, The Latin Translator for the translations of the Latin quotes in the chapter by Ernst Mach.

Reginald Smith, Proud Member
Citizen Scientists League
May 22, 2012

About the Citizen Scientists League

http://citizenscientistsleague.com

America is, and has always been, a nation of the curious, the seeker, the inventor. Those scientific luminaries among its Founding Fathers such as Benjamin Franklin and Thomas Jefferson were deeply committed to the Enlightenment ideal that people should have the latitude to seek out their own answers to questions, unfettered by dogmas of any kind. They were, moreover, committed to the quest of knowledge for its own sake. The Citizen Scientists League is a nonprofit organization that traces its roots from this tradition, but more recently through the great amateur scientists and independent thinkers of more recent decades; Albert Ingalls, C.L. Stong, Jearl Walker, Forrest Mims, Shawn Carlson, Martin Gardner, Bucky Fuller, and many others.

The Citizen Scientists League enables people of all ages and levels of experience or expertise to participate in the adventure of scientific discovery by creating a community of the curious, where ideas and instruction can flow and mingle to help create new generations of intellectual audacity and uncommon ingenuity.

Mission Statement:

The Citizen Scientists League promotes responsible scientific observation, experimentation, discovery, and invention. We encourage active participation, networking and publishing by science enthusiasts at all levels of education and experience.

PART I
Starting out in Research

1

Advice to a Young Investigator: Chapters 1 & 2 Santiago Ramón y Cajal

Santiago Ramón y Cajal (May 1, 1852 – October, 17, 1934)—was a Spanish pathologist who is considered one of the founders of neuroscience for his detailed research on the structure of the brain. He won the Nobel Prize in Physiology or Medicine in 1906 for his research. This text selection is from the first two chapters of his book Advice to a Young Investigator which is a valuable compendium of advice for researchers on scientific work and creativity.

CHAPTER 1: Introduction

Thoughts about general methods. Abstract rules are sterile. Need to enlighten the mind and strengthen resolve. Organization of the book

I shall assume that the reader's general education and background in philosophy are sufficient to understand that the major sources of knowledge include observation, experiment, and reasoning by induction and deduction.

Instead of elaborating on accepted principles, let us simply point out that for the last hundred years the natural sciences have abandoned completely the Aristotelian principles of intuition, inspiration, and dogmatism.

The unique method of reflection indulged in by the Pythagoreans and followers of Plato (and pursued in modern times by Descartes, Fichte, Krause, Hegel, and more recently at least partly by Bergson) involves exploring one's own mind or soul to discover universal laws and solutions to the great secrets of life. Today this approach can only generate feelings of sorrow and compassion—the latter because of talent wasted in the pursuit of chime-

---

* Reprinted with permission from Santiago Ramón y Cajal, *Advice to a Young Investigator* (Cambridge, MA: MIT Press, 1999) Translated by Neely Swanson and Larry W. Swanson

ras, and the former because of all the time and work so pitifully squandered.

The history of civilization proves beyond doubt just how sterile the repeated attempts of metaphysics to guess at nature's laws have been. Instead, there is every reason to believe that when the human intellect ignores reality and concentrates within, it can no longer explain the simplest inner workings of life's machinery or of the world around us.

The intellect is presented with phenomena marching in review before the sensory organs. It can be truly useful and productive only when limiting itself to the modest tasks of observation, description, and comparison, and of classification that is based on analogies and differences. A knowledge of underlying causes and empirical laws will then come slowly through the use of inductive methods. Another commonplace worth repeating is that science cannot hope to solve Ultimate Causes. In other words, science can never understand the foundation hidden below the appearance of phenomena in the universe. As Claude Bernard has pointed out, researchers cannot transcend the determinism of phenomena; instead, their mission is limited to demonstrating the *how*, never the *why*, of observed changes. This is a modest goal in the eyes of philosophy, yet an imposing challenge in actual practice. Knowing the conditions under which a phenomenon occurs allows us to reproduce or eliminate it at will, therefore allowing us to control and use it for the benefit of humanity. Foresight and action are the advantages we obtain from a deterministic view of phenomena.

The severe constraints imposed by determinism may appear to limit philosophy in a rather arbitrary way [1]. However, there is no denying that in the natural sciences—and especially in biology—it is a very effective tool for avoiding the innate tendency to explain the universe as a whole in terms of general laws. They are like a germ with all the necessary parts, just as a seed contains all the potentialities of the future tree within it. Now and then philosophers invade the field of biological sciences with these beguiling generalizations, which tend to be unproductive, purely verbal solutions lacking in substance. At best, they may prove useful when viewed simply as working hypotheses.

Thus, we are forced to concede that the "great enigmas" of the universe listed by Du Bois-Raymond are beyond our understanding at the present time. The great German physiologist pointed out that we must resign ourselves to the state of *ignoramus*, or even the inexorable *ignorabimus*.

There is no doubt that the human mind is fundamentally incapable of solving these formidable problems (the origin of life, nature of matter, origin of movement, and appearance of consciousness). Our brain is an organ of action that is directed toward practical tasks; it does not appear to have been built for discovering the ultimate causes of things, but rather for determining their immediate causes and invariant relationships. And whereas this may appear to be very little, it is in fact a great deal. Having been granted the immense advantage of participating in the unfolding of our world, and of modifying it to life's advantage, we may proceed quite nicely without knowing the essence of things.

It would not be wise in discussing general principles of research to overlook those panaceas of scientific method so highly recommended by Claude Bernard, which are to be found in Bacon's *Novum Organum* and Descartes's *Book of Methods*. They are exceptionally good at stimulating thought, but are much less effective in teaching one how to discover. After confessing that reading them may suggest fruitful idea or two, I must further confess an inclination to share De Maistre's view of the *Novum Organum*: "Those who have made the greatest discoveries in science never read it, and Bacon himself failed to make a single discovery based on his own rules." Liebig appears even more harsh in his celebrated *Academic Discourse* when he states that Bacon was a scientific dilettante whose writings contain nothing of the processes leading to discovery, regardless of inflated praise from jurists, historians, and others far removed from science.

No one fails to use instinctively the following general principles of Descartes when approaching any difficult problem: "Do not acknowledge as true anything that is not obvious, divide a problem into as many parts as necessary to attack it in the best way, and start an analysis by examining the simplest and most easily understood parts before ascending gradually to an understanding of the most complex." The merit of the French philosopher is not based on his application of these principles but rather on having formulated them clearly and rigorously after having profited by them unconsciously, like everyone else, in his thinking about philosophy and geometry.

I believe that the slight advantage gained from reading such work, and in general any work concerned with philosophical methods of investigation, is based on the vague, general nature of the rules they express. In other words, when they are not simply empty formulas they become formal expressions of the mechanism of understanding used during the process of research. This mechanism acts unconsciously in every well-organized and cultivated mind, and when the philosopher reflexly formulates psychological principles, neither the author nor the reader can improve their respective abilities for scientific investigation. Those writing on logical methods impress me in the same way as would a speaker attempting to improve his eloquence by learning about brain speech centers, about voice mechanics, and about the distribution of nerves to the larynx—as if knowing these anatomical and physiological details would create organization where none exists, or refine what we already have [2].

It is important to note that the most brilliant discoveries have not relied on a formal knowledge of logic. Instead, their discoverers have had an acute inner logic that generates ideas with the same unstudied unconsciousness that allowed Jourdain to create prose. Reading the work of the great scientific pioneers such as Galileo, Kepler, Newton, Lavoisier, Geoffroy Saint-Hilaire, Faraday, Ampere, Bernard, Pasteur, Virchow, and Liebig is considerably more effective. However, it is important to realize that if we lack even a spark of the splendid light that shone in those minds, and at least a trace of the noble zeal that motivated such distinguished individuals, this exercise may if nothing else convert us to enthusiastic or insightful commentators on their work—perhaps even to good scientific writers—but it will not create the spirit of investigation within us.

A knowledge of principles governing the historical unfolding of science also provides no great advantage in understanding the process of research. Herbert Spencer proposed that intellectual progress emerges from that which is homogeneous and that which is heterogeneous, and by virtue of the *instability of that which is homogeneous*, and of the principle that *every cause produces more than one effect*, each discovery immediately stimulates many other discoveries. However, even if this concept allows us to appreciate the historical march of science, it cannot provide us with the key to its revelations. The important thing is to discover how each investigator, in his own special domain, was able to segregate heterogeneous from homogeneous, and to learn why many of those who set out to accomplish a particular goal did not succeed.

Let me assert without further ado that there are no rules of logic for making discoveries, let alone for converting those lacking a natural talent for thinking logically into successful researchers. As for geniuses, it is well-known that they have difficulty bowing to rules—they prefer to make them instead. Condorcet has noted that "The mediocre can be educated; geniuses educate themselves."

Must we therefore abandon any attempt to instruct and educate about the process of scientific research? Shall we leave the beginner to his own devices, confused and abandoned, struggling without guidance or advice along a path strewn with difficulties and dangers?

Definitely not. In fact, just the opposite—we believe that by abandoning the ethereal realm of philosophical principles and abstract methods we can descend to the solid ground of experimental science, as well as to the sphere of ethical considerations involved in the process of inquiry. In taking this course, simple, genuinely useful advice for the novice can be found.

In my view, some advice about what should be known, about what technical education should be acquired, about the intense motivation needed to succeed, and about the carelessness and inclination toward bias that must be avoided, is far more useful than all the rules and warnings of theoretical logic. This is the justification for the present work, which contains those encouraging words and paternal admonitions that the writer would have liked so much to receive at the beginning of his own modest scientific career.

My remarks will not be of much value to those having had the good fortune to receive an education in the laboratory of a distinguished scientist, under the beneficial influence of living rules embodied in a learned personality who is inspired by the noble vocation of science combined with teaching. They will also be of little use to those energetic individuals—those gifted souls mentioned above—who obviously need only the guidance provided by study and reflection to gain an understanding of the truth. Nevertheless, it is perhaps worth repeating that they may prove comforting and useful to the large number of modest individuals with a retiring nature who, despite yearning for reputation, have not yet reaped the desired harvest, due either to a certain lack of determination or to misdirected efforts.

This advice is aimed more at the spirit than the intellect because I am convinced, and Payot wisely agrees, that the former is as amenable to education as the latter. Furthermore, I believe that all outstanding work, in art as well as in science, results from immense zeal applied to a great idea. The present work is divided into nine chapters. In the second I will try to show how the prejudices and lax judgment that weaken the novice can be avoided. These problems destroy the self-confidence needed for any investigation to reach a happy conclusion. In the third chapter I will consider the moral values that should be displayed—which are like stimulants of the will. In the fourth chapter I will suggest what needs to be known in preparing for a competent struggle with nature. In the fifth, I will point out certain impairments of the will and of judgment that must be avoided. In the sixth, I will discuss social conditions that favor scientific work, as well as influences of the family circle. In the seventh, I will outline how to plan and carry out the investigation itself (based on observation, explanation or hypothesis, and proof). In the eighth I will deal with how to write scientific papers; and finally, in the ninth chapter the investigator's moral obligations as a teacher will be considered.

Notes

[1] In attempting to prove his hypothesis, Claude Bernard may have exaggerated somewhat in claiming that: "We shall never know why opium has soporific effects, or why the combination of

hydrogen and oxygen yields a substance so different in physical and chemical properties as water." The impossibility of reducing the properties of matter to laws governing the position, form, and movement of atoms (today we would say of ions and electrons) seems real at this time, but it does not seem that it should be thus in principle and forever. (Author's footnote, 1923.)

[2] It is extraordinary how well this theory agrees with one elaborated by Schopenhauer (which was unknown to us at the time this essay was first published) in his book *The World as Will and as Representation*, pp. 98 ff. Concerning logic, he says that "the best logic for a particular science abandons the rules of logic when it begins serious discourse." And further on: "Wanting to make practical use of logic is like consulting the field of mechanics before learning to walk." More recently, Eucken expressed a similar view in saying that "rules and forms of logic are not enough to produce an ingenious thought." (Author's footnote, 1923.)

CHAPTER 2: Beginner's Traps

Undue admiration of authority. The most important problems are already solved. Preoccupation with applied science. Perceived lack of ability

Undue admiration of authority

I believe that excessive admiration for the work of great minds is one of the most unfortunate preoccupations of intellectual youth—along with a conviction that certain problems cannot be attacked, let alone solved, because of one's relatively limited abilities.

Inordinate respect for genius is based on a commendable sense of fairness and modesty that is difficult to censure. However, when foremost in the mind of a novice, it cripples initiative and prevents the formulation of original work. Defect for defect, arrogance is preferable to diffidence, boldness measures its strengths and conquers or is conquered, and undue modesty flees from battle, condemned to shameful inactivity.

When one escapes the atmosphere of stylistic legerdemain inhaled while reading the published work of a genius, and enters the laboratory to confirm the observations upon which the intriguing ideas are based, now and then hero worship declines as self-esteem grows. Great men are at times geniuses, occasionally children, and always incomplete. Even when the work of a genius is subjected to critical analysis and no errors are found, it is important to realize that everything he has discovered in a particular field is almost nothing in comparison with what remains to be discovered. Nature offers inexhaustible wealth to all. There is certainly no reason to envy our predecessors, or to exclaim with Alexander following the victories of Philip, "My father is going to leave me nothing to conquer!"

Admittedly, certain concepts in science appear to be so complete, brilliant, and enduring that they seem to be the fruit of an almost divine intuition, springing forth perfect like Minerva from the head of Jupiter. However, the well-deserved admiration for such accomplishments would be considerably diminished were we aware of all of the time and effort, patience and perseverance, trials, corrections, and even mishaps that worked hand in hand to produce the final success—contributing almost as much as the investigator's genius. The same principle applies to the marvelous adaptation of the human organism to predetermined functions. When examined alone, the vertebrate eye or ear is a source of amazement. It seems impossible that these organs could have formed simply by the collective action of natural laws.

However, when we consider all of the gradations and transitional forms that they display in the phylogenetic series, from the almost shapeless ocular outline of certain infusoria and worms to the complicated organization of the eye in lower vertebrates, not one whit of our admiration is lost and our minds are apt to accept the idea of natural formation through the mechanisms of variation, organic correlation, natural selection, and adaptation [1].

What a wonderful stimulant it would be for the beginner if his instructor, instead of amazing and dismaying him with the sublimity of great past achievements, would reveal instead the origin of each scientific discovery, the series of errors and missteps that preceded it—information that, from a human perspective, is essential to an accurate explanation of the discovery. Skillful pedagogical tactics such as this would instill the conviction that the discoverer, along with being an illustrious person of great talent and resolve, was in the final analysis a human being just like everyone else.

Far from humbling one's self before the great authorities of science, those beginning research must understand that—by a cruel but inevitable law—their destiny is to grow a little at the expense of the great one's reputation. It is very common for those beginning their scientific explorations with some success to do so by weakening the pedestal of an historic or contemporary hero. By way of classic examples, recall Galileo refuting Aristotle's view of gravity, Copernicus tearing down Ptolemy's system of the universe, Lavoisier destroying Stahl's concept of phlogiston, and Virchow refuting the idea of spontaneous generation held by Schwann, Schleiden, and Robin. This principle is so general and compelling that it is displayed in all areas of science and extends to even the humblest of investigators. If I might be so bold as to refer to myself in the company of such eminent examples, I should add that on initiating my own work on the anatomy and physiology of nervous centers, the first obstacle that had to be set aside was the false theory of Gerlach and Golgi on the diffuse nature of neural networks in the gray matter, and on the nature of nerve current transmission.

Two phases may often be noted in the careers of learned investigators. First there is the productive time devoted to the elimination of past errors and the illumination of new data, and it is followed by the mature or intellectual phase (which does not necessarily coincide with old age) when scientific productivity declines and the hypotheses incubated during youth are defended with paternal affection from the attacks of newcomers [2]. Throughout history, no great man has shunned titles or failed to extol his right to glory before the new generation. Rousseau's bitter quote is sad but true: "There has never been a wise man who hasn't failed to prefer the lie invented by himself to the truth discovered by someone else."

Even in the most exact sciences there are always some laws that are maintained exclusively through the force of authority. To demonstrate their inaccuracy with new research is always an excellent way to begin genuine scientific work. It hardly matters whether the correction is received with harsh criticism, traitorous invective, or silence, which is even more cruel. Because right is on his side, the innovator will quickly attract the young, who obviously have no past to defend. And those impartial scholars who, in the midst of the smothering torrent of current doctrine, have learned how to keep their minds clear and their judgment independent will also gather on his side.

However, it is not enough to destroy—one must also build. Scientific criticism is justified only by establishing truth in place of error. Generally speaking, new principles emerge from the ruins of those abandoned, based strictly on facts correctly interpreted. The innovator must avoid all pious concessions to traditional error and crumbling ideas if he does not wish to see his fame quickly shared by the critics and those merely focusing on details, who immediately sprout in great numbers after each discovery, like mushrooms in the shade of a tree.

The Most Important Problems Are Already Solved

Here is another false concept often heard from the lips of the newly graduated: "Everything of major importance in the various areas of science has already been clarified. What difference does it make if I add some minor detail or gather up what is left in some field where more diligent observers have already collected the abundant, ripe grain. Science won't change its perspective because of my work, and my name will never emerge from obscurity."

This is often indolence masquerading as modesty. However, it is also expressed by worthy young men reflecting on the first pangs of dismay experienced when undertaking some major project. This superficial concept of science must be eradicated by the young investigator who does not wish to fail, hopelessly overcome by the struggle developing in his mind between the utilitarian suggestions that are part and parcel of his ethical environment (which may soon convert him to an ordinary and financially successful general practitioner), and those nobler impulses of duty and loyalty urging him on to achievement and honor.

Wanting to earn the trust placed in him by his mentors, the inexperienced observer hopes to discover a new lode at the earth's surface, where easy exploration will build his reputation quickly.

Unfortunately, with his first excursions into the literature hardly begun, he is shocked to find that the metal lies deep within the ground—surface deposits have been virtually exhausted by observers fortunate enough to arrive earlier and exercise their simple right of eminent domain.

It is nevertheless true that if we arrived on the scene too late for certain problems, we were also born too early to help solve others. Within a century we shall come, by the natural course of events, to monopolize science, plunder its major assets, and harvest its vast fields of data.

Yet we must recognize that there are times when, on the heels of a chance discovery or the development of an important new technique, magnificent scientific discoveries occur one after another as if by spontaneous generation. This happened during the Renaissance when Descartes, Pascal, Galileo, Bacon, Boyle, Newton, our own Sanchez, and others revealed clearly the errors of the ancients and spread the belief that the Greeks, far from exhausting the field of science, had scarcely taken the first steps in understanding the universe [3]. It is a wonderful and fortunate thing for a scientist to be born during one of these great decisive moments in the history of ideas, when much of what has been done in the past is invalidated. Under these circumstances, it could not be easier to choose a fertile area of investigation.

However, let us not exaggerate the importance of such events. Instead, bear in mind that even in our own time science is often built on the ruins of theories once thought to be indestructible. It is important to realize that if certain areas of science appear to be quite mature, others are in the process of development, and yet others remain to be born. Especially in biology, where immense amounts of work have been carried out during the last century, the most essential problems remain unsolved—the origin of life, the problems of heredity and development, the structure and chemical composition of the cell, and so on.

It is fair to say that, in general, no problems have been exhausted; instead, men have been exhausted by the problems. Soil that appears impoverished to one researcher reveals its fertility to another. Fresh talent approaching the analysis of a problem without prejudice will always see new possibilities—some aspect not considered by those who believe that a subject is fully understood. Our knowledge is so fragmentary that unexpected findings appear in even the most fully explored topics. Who, a few short years ago, would have suspected that light and heat still held scientific secrets in reserve? Nevertheless, we now have *argon* in the atmosphere, the *x-rays* of Roentgen, and the *radium* of the Curies, all of which illustrate the inadequacy of our former methods, and the prematurity of our former syntheses.

The best application of the following beautiful dictum of Geoffroy Saint-Hilaire is in biology: "The infinite is always before us." And the same applies to Carnoy's no less graphic thought: "Science is a perpetual creative process." Not everyone is destined to venture into the forest and by sheer determination carve out a serviceable road. However, even the most humble among us can take advantage of the path opened by genius and by traveling along it extract one or another secret from the unknown.

If the beginner is willing to accept the role of gathering details that escaped the wise discoverer, he can be assured that those searching for minutiae eventually acquire an analytical sense so discriminating, and powers of observation so keen, that they are able to solve important problems successfully.

So many apparently trivial observations have led investigators with a thorough knowledge of methods to great scientific conquests! Furthermore, we must bear in mind that because science relentlessly differentiates, the minutiae of today often become important principles tomorrow.

It is also essential to remember that our appreciation of what is important and what is minor, what is great and what is small, is based on false wisdom, on a true anthropomorphic error. Superior and inferior do not exist in nature, nor do primary and secondary relationships. The hierarchies that our minds take pleasure in assigning to natural phenomena arise from the fact that instead of considering things individually, and how they are interrelated, we view them strictly from the perspective of their usefulness or the pleasure they give us. In the chain of life all links are equally valuable because all prove equally necessary. Things that we see from a distance or do not know how to evaluate are considered small.

Even assuming the perspective of human egotism, think how many issues of profound importance to humanity lie within the protoplasm of the simplest microbe! Nothing seems more important in bacteriology than a knowledge of infectious bacteria, and nothing more secondary than the inoffensive microbes that grow abundantly in decomposing organic material. Nevertheless, if these humble fungi—whose mission is to return to the general circulation of matter those substances incorporated by the higher plants and animals—were to disappear, humans could not inhabit the planet.

The far-reaching importance of attention to detail in technical methodology is perhaps demonstrated more clearly in biology than in any other sphere. To cite but one example, recall that Koch, the great German bacteriologist, thought of adding a little alkali to a basic aniline dye, and this allowed him to stain and thus discover the tubercle bacillus—revealing the etiology of a disease that had until then remained uncontrolled by the wisdom of the most illustrious pathologists.

Even the most prominent of the great geniuses have demonstrated a lack of intellectual perspective in the appraisal of scientific insights. Today, we can find many seeds of great discoveries that were mentioned as curiosities of little importance in the writings of the ancients, and even in those of the wise men of the Renaissance. Lost in the pages of a confused theological treatise (*Christianismi restitutio*) are three apparently disdainful lines written by Servetus referring to the pulmonary circulation, which now constitute his major claim to fame. The Aragonese philosopher would be surprised indeed if he were to rise from the dead today. He would find his laborious metaphysical disquisitions totally forgotten, whereas the observation he used simply to argue for the residence of the soul in the blood is widely praised! Or again, it has been inferred from a passage of Seneca's that the ancients knew the magnifying powers of a crystal sphere filled with water. Who would have suspected that in this phenomenon of magnification, disregarded for centuries, slumbered the embryo of two powerful analytical instruments, the microscope and telescope—and two equally great sciences, biology and astronomy!

In summary, there are no small problems. Problems that appear small are large problems that are not understood. Instead of tiny details unworthy of the intellectual, we have men whose tiny intellects cannot rise to penetrate the infinitesimal. Nature is a harmonious mechanism where all parts, including those appearing to play a secondary role, cooperate in the functional whole. In contemplating this mechanism, shallow men arbitrarily divide its parts into essential and secondary, whereas the insightful thinker is content with classifying them as understood and poorly understood, ignoring for the moment their size and immediately useful properties. No one can predict their importance in the future.

Preoccupation with Applied Science

Another corruption of thought that is important to battle at all costs is the false distinction between *theoretical* and *applied* science, with accompanying praise of the latter and deprecation of the former. This error spreads unconsciously among the young, diverting them from the course of disinterested inquiry.

This lack of appreciation is definitely shared by the average citizen, often including lawyers, writers, industrialists, and unfortunately even distinguished statesmen, whose initiatives can have serious consequences for the cultural development of their nation. They should avoid expressing the following sentiments:

> *Fewer doctors and more industrialists. The greatness of nations is not measured by what the former know, but rather by the number of scientific triumphs applied to commerce, industry, agriculture, medicine, and the military arts. We shall leave to the phlegmatic and lazy Teutons their subtle investigations of pure science and mad eagerness to pry into the remotest corners of life. Let us devote ourselves to extracting the practical essence of scientific knowledge, and then using it to improve the human condition. Spain needs machines for its trains and ships, practical advances for agriculture and industry, a rational health care system—in short, whatever contributes to the common good, the nation's wealth, and the people's well-being. May God deliver us from worthless scholars immersed in dubious speculation or dedicated to the conquest of the infinitesimal, which would be considered a frivolous if not ridiculous pastime if it weren't so expensive.*

Ineptitudes like this are formulated at every step by those who, while traveling abroad, see progress as a strange mirage of effects rather than causes. People with little understanding fail to observe the mysterious threads that bind the factory to the laboratory, just as the stream is connected with its source. Like the man in the street, they believe in good faith that scholars may be divided into two groups—those who waste time speculating about unfruitful lines of pure science, and those who know how to find data that can be applied immediately to the advancement and comfort of life [4].

Is it really necessary to dwell on such an absurd point of view? Does anyone lack the common sense to understand that applications derive immediately from the discovery of fundamental principles and new data? In Germany, France, and England the factory and laboratory are closely intertwined, and very often the scientist himself (either personally or through a development company) directs its industrial application. Such alliances are obvious in the great aniline dye factories that are one of the richest lodes of German, Swiss, and French industry. This is so well known that examples are hardly necessary. Nevertheless, I would like to cite two recent developments that are very significant.

One is the great industry involved in the manufacture of precision lenses (for micrography, photography, and astronomy). It was created in Germany by the profound work in mathematical optics of Professor Abbé of Jena, and it gives Prussia an enormously valuable monopoly that is supported by the entire world [5]. The other example is the manufacture of therapeutic serums that was born in Berlin and perfected in Paris. It is both natural and legitimate that Behring and Roux, who established the scientific principles upon which serum therapy is based, exercise a controlling hand.

For the present, let us cultivate science for its own sake, without considering its applications. They will always come, whether in years or perhaps even in centuries. It matters very little whether scientific truth is used by our sons or by our grandsons. The course of progress obviously would have suffered if Galvani, Volta, Faraday, and Hertz, who discovered the fundamental principles of electricity, had discounted their findings because there were no industrial applications for them at the time. Accept the view that nothing in nature is useless, even from the human point of view (with the necessary restrictions of time and place). Even in the rare instance where it may not be possible to use particular scientific breakthroughs for our comfort and benefit, there is one positive benefit—the noble satisfaction of our curiosity and the incomparable gratification and feeling of power that accompany the solving of a difficult problem.

In short, consider problems on their own merits when attacking them. Avoid deviating to secondary concerns that distract attention and weaken analytical powers. In struggling with nature, the biologist, like the astronomer, must look beyond the earth he lives on and concentrate on the serene universe of ideas, where the light of truth will eventually shine. The applications of new data will come in due time, when other related information emerges. It is well known that a discovery is simply the joining of two or more pieces of information to a useful end. Many scientific observations are of little use at the time they are made. However, after some decades, or perhaps even centuries, a new discovery clarifies the old, and the resulting *industrial application* may be called photography, the phonograph, spectral analysis, wireless telegraphy, or mechanical flight. Synthesis occurring over a variable length of time is always involved. Porta discovered the principle of the *camera obscura*, an isolated event that had very little impact on the art of design. Wedgwood and Davy noted in 1802 the possibility of obtaining photographic images on a certain type of paper immersed in silver nitrate solution, but this had little impact because the copy could not be fixed. Then came John Herschel, who succeeded in dissolving the silver salt not affected by light, and with this it was possible to fix the fugitive luminous silhouette. However, despite this advance, Porta's apparatus was virtually impossible to use because the silver salts available at the time were so weak. Then Daguerre finally appeared. He discovered the latent image in 1839 by using the much greater sensitivity of silver iodide. Daguerre admirably synthesized the inventions of his predecessors and used the foundation that they laid to create the science of photography as we know it today.

All inventions evolve in this way. Information is transmitted through time by discerning though unlucky observers who fail to harvest the fruits of their labor, which await fertilization. Nevertheless, once data are gathered, a scientist will come along at some point who is fortunate not so much for his originality as for having been born at an opportune moment. He considers the facts from the human point of view, synthesizes, and a discovery emerges.

Perceived Lack of Ability

Some people claim a lack of ability for science to justify failure and discouragement. "I enjoy laboratory work," they tell us, "but am no good at discovering things." Certainly there are minds unsuited for experimental work, especially if they have a short attention span and lack curiosity and admiration for the work of nature. But are the great majority of those professing incompetence really so? Might they exaggerate how difficult the task will be, and underestimate their own abilities? I believe that this is often the case, and would even venture to suggest that many people habitually confuse inability with the simple fact that they learn and understand slowly, or perhaps are sometimes even lazy or they don't have a secondary trait such as patience, thoroughness, or determination—which may be acquired rapidly through hard work and the satisfaction of success.

In my opinion the list of those suited for scientific work is much longer than generally thought, and contains more than the superior talents, readily adaptable, and keen minds ambitious for reputation and eager to link their names with a major discovery. The list also includes those ordinary intellects thought of as *skillful* because of the ability and steadiness they display for all manual work, those gifted with artistic talent who appreciate deeply the beauty of Nature's work, and those who are simply curious, calm, and phlegmatic devotees of the religion of detail, willing to dedicate long hours to examining the most insignificant natural phenomena. Science, like an army, needs generals as well as soldiers; plans are conceived by the former, but the latter actually conquer. Merely through being less brilliant, the collaboration of those who perfect and carry out the original plan cannot fail to be highly valuable. Thanks to these workers in the march of progress, the concept of a genius acquires vigor and clarity, transformed from abstract symbol to live reality, appreciated and known by all.

Various procedures can be used to assess one's aptitude for laboratory work. Based on my experience, I would recommend the following two:

1. Attempt to repeat some analytical method that is considered unreliable and difficult until patience and hard work yield results similar to those published by the author. Pleasure derived from success, especially if it has come without the supervision of an instructor (that is, working alone), is a clear indication of aptitude for experimental work.
2. Find a scientific topic that is difficult and surrounded by controversy, and examine it superficially by reading general reference books instead of detailed monographs. Then, after several months of experimental work, our beginner should consult the latest literature on the subject. If he has arrived at similar conclusions, if his thinking on hotly disputed points falls in line with the interpretations of noted authorities, and if he has managed to avoid the errors committed by certain authors, then timidity should be abandoned, and scientific work should be approached without reservation. Many triumphs and satisfactions lie ahead, depending on how hard one works.

Even those with modest intellectual abilities will gather some fruit, provided they maintain faith in the creative power of education and devote extended periods of time to thorough analysis of a focused topic.

At the risk of appearing repetitive, tiresome, and boring, I would like to present the following reflections to counter those who do not believe in the power of determination. As many teachers and thinkers have noted, discoveries are not the fruit of outstanding talent, but rather of common sense enhanced and strengthened by technical education and a habit of thinking about scientific problems [6]. Thus, anyone with mental gifts balanced enough to cope with everyday life may use them to progress successfully along the road of investigation.

The youthful brain is wonderfully pliable and, stimulated by the *impulses of a strong will to do so*, can greatly improve its organization by creating new associations between ideas and by refining the powers of judgment.

Deficiencies of innate ability may be compensated for through persistent hard work and concentration. One might say that work substitutes for talent, or better yet that it *creates talent.* He who firmly determines to improve his capacity will do so, provided that education does not begin too late, during a period when the plasticity of nerve cells is greatly reduced. Do not forget that reading and thinking about masterpieces allows one to assimilate much of the skill that created them, providing of course that one extends beyond conclusions to the author's insights, guiding principles, and even style.

What we refer to as a great and special talent usually implies superiority that is *expeditious* rather than *qualitative.* In other words, it simply means doing quickly and with brilliant success what ordinary intellects carry out slowly but well. Instead of distinguishing between mediocre and great minds, it would be preferable and more correct in most instances to classify them as *slow* and *facile* [7]. The latter are certainly more brilliant and stimulating—there is no substitute for them in conversation, oratory, and journalism, that is, in all lines of work where time is a decisive factor. However, in scientific undertakings the *slow* prove to be as useful as the fast because scientists like artists are judged by the quality of what they produce, not by the speed of production. I would even venture to add that as a very common compensation *slow* brains have great endurance for prolonged concentration. They open wide, deep furrows in problems, whereas facile brains often tire quickly after scarcely clearing the land. There are, however, many exceptions to this generalization: Newton, Davy, Pasteur, Virchow, and others were active minds who left a broad, luminous wake.

If our memory is inconsistent and weak, despite efforts to improve, then let us *manage it well.* As Epictetus said: "When you are dealt poor cards in the game of life, there is nothing to do but make the best of them." History teaches of the occasional great discoveries made by those with ordinary minds and memory ably used, rather than by those with superior abilities. Great scientific innovators such as Helmholtz have complained of bad memory—of how learning prose by rote is akin to torture! As compensation, those with short memories for words and phrases seem to enjoy excellent retention of ideas and logical arguments. And Locke has pointed out that those endowed with great genius and a facile memory do not always excel in judgment.

To pursue fully the topic of our research with the limited facilities that we have, let us forget unrelated pursuits and the parasitic ideas connected with the futile trifles of everyday life. Using strength and perseverance, concentrate deeply only on information pertinent to the question at hand. During the gestation period of our work, sentence ourselves to ignorance of everything else that is going on—politics, literature, music, and idle gossip. There are occasions when ignorance is a great virtue, almost a state of heroism. Useless books distract attention and are thus weighty, occupying as much space in our brains as on the library shelf. They can spoil or hinder mental adjustments to the problem at hand. Although popular opinion may not agree, "Knowledge occupies space."

Even those with mediocre talent can produce notable work in the various sciences, so long as they do not try to embrace all of them at once. Instead, they should concentrate attention on one subject after another (that is, in different periods of time), although later work will undermine earlier attainments in the other spheres. This amounts to saying that the brain adapts to universal science in *time* but not in *space*. In fact, even those with great abilities proceed in this way. Thus, when we are astonished by someone with publications in different scientific fields, realize that each topic was explored during a specific period of time. Knowledge gained earlier certainly will not have disappeared from the mind of the author, but it will have become simplified by condensing into formulas or greatly abbreviated symbols. Thus, ample space remains for the perception and learning of new images on the cerebral blackboard.

Notes

[1] I believe less in the power of natural selection today than I did when I wrote these lines in 1893. The more I study the organization of the eye in vertebrates and invertebrates, the less I understand the causes of their marvelous and exquisitely adapted organization.

[2] Ostwald corroborates this view in a recent book, noting that almost all the great discoveries have been the work of youth. Newton, Davy, Faraday, Hertz, and Mayer are good examples.

[3] The brilliant series of discoveries in electricity that followed Volta's development of the voltaic pile at the beginning of the last century, the Pleiades of histological work inspired by Schwann's discovery of cell multiplication, and the profound repercussions that the not so distant finding of roentgen rays have produced in all areas of physics (the observation of radioactivity, and the discovery of radium and polonium and of the phenomenon of emanation) are good examples of that creative and, in a sense, automatic virtue possessed by all great discoveries, which seem to grow and multiply like seeds cast by chance on fertile soil.

[4] This popular view has been refuted eloquently by many scholars. However, I can't resist the temptation to quote a comparison that has been made in various brilliant forms, here by our incomparable scientific commentator, José Echegaray, who did so much to translate science into popular terms, and whose death robbed Spanish science of a great talent: Pure science is like a beautiful cloud of gold and scarlet that diffuses wondrous hues and beams of light in the west. It is not an illusion, but the splendor and beauty of truth. However, now the cloud rises, the winds blow it over the fields, and it takes on darker, more somber colors. It is performing a task and changing its party clothes—think of it as putting on its work shirt. It generates rain that irrigates the fields, soaking the land and preparing it for future harvests. In the end it provides humanity with its daily bread. What began as beauty for the soul and intellect ends by providing nourishment for the humble life of the body. Academia de Sciencias, formal session of March 12, 1916.

[5] This was written in 1896. Now [1923] there are no fewer than thirty-three outstanding researchers in mathematics, optics, mechanics, and chemistry at the optical instruments factory in Jena. Furthermore, legions of chemists also work in the great German factories that produce chemical products. It is clear that the only way for industry to avoid routine and stagnation is to convert the laboratory to an antechamber of the factory.

[6] "It is common sense to work under considerable stress," according to the graphic adage of Echegaray.

[7] This view is consistent with the classification of classic and romantic (applied to minds that react slowly and minds that react quickly) provided by Ostwald in his interesting recent book, *Great Men* (Ed.—in German *Grosse Manner*).

2

The Art of Scientific Investigation: Chapter 1: Preparations
William Ian Beardmore Beveridge

William Ian Beardmore (W.I.B.) Beveridge (April 23, 1908 – August 14, 2006)—was an Australian microbiologist and animal pathologist at Cambridge. He is perhaps best known for his work with future Nobel Laureate Frank Burnet (Medicine & Physiology 1960) in devising a technique for cultivating viruses on the chorioallantoic membrane of chicken embryos. He later headed the Department of Animal Pathology at Cambridge. In his book The Art of Scientific Investigation, which will have four excerpts in this book, he discusses the techniques of creative problem solving and fruitful scientific research. In chapter 1, he discusses first preparations for research.

"The lame in the path outstrip the swift who wander from it."
—Francis Bacon

Study

The research worker remains a student all his life. Preparation for his work is never finished for he has to keep abreast with the growth of knowledge. This he does mainly by reading current scientific periodicals. Like reading the newspapers, this study becomes a habit and forms a regular part of the scientist's life.

The 1952 edition of the *World List of Scientific Periodicals* indexes more than 50,000 periodicals. A simple calculation shows this is equivalent to probably two million articles a year, or 40,000 a week, which reveals the utter impossibility of keeping abreast of more than the small fraction of the literature which is most perti-

* Reprinted from W.I.B. Beveridge, *The Art of Scientific Investigation* (New York: W.W. Norton, 1957) 1-8 under Creative Commons Attribution-ShareAlike License 3.0

nent to one's interest. Most research workers try to see regularly and at least glance through the titles of the articles in twenty to forty periodicals. As with the newspaper, they just skim through most of the material and read fully only those articles which may be of interest.

The beginner would be well advised to ask an experienced research worker in his field which journals are the most important for him to read. Abstracting journals are of limited value, if only because they necessarily lag some considerable time behind the original journals. They do, however, enable the scientist to cover a wide range of literature and are most valuable to those who have not access to a large number of journals. Students need some guidance in ways of tracing references through indexing journals and catalogues and in using libraries.

It is usual to study closely the literature dealing with the particular problem on which one is going to work. However, surprising as it may seem at first, some scientists consider that this is unwise. They contend that reading what others have written on the subject conditions the mind to see the problem in the same way and makes it more difficult to find a new and fruitful approach. There are even some grounds for discouraging an excessive amount of reading in the general field of science in which one is going to work. Charles Kettering, who was associated with the discovery of tetraethyl lead as an anti-knock agent in motor fuels and the development of diesel engines usable in trucks and buses, said that from studying conventional text-books we fall into a rut and to escape from this takes as much effort as to solve the problem. Many successful investigators were not trained in the branch of science in which they made their most brilliant discoveries: Pasteur, Metchnikoff and Galvani are well-known examples. A sheepman named J.H.W. Mules, who had no scientific training, discovered a means of preventing blowfly attack in sheep in Australia when many scientists had failed. Bessemer, the discoverer of the method of producing cheap steel, said:

> *I had an immense advantage over many others dealing with the problem inasmuch as I had no fixed ideas derived from long established practice to control and bias my mind, and did not suffer from the general belief that whatever is, is right.*

But in his case, as with many such "outsiders," ignorance and freedom from established patterns of thought in one field were joined with knowledge and training in other fields. In the same vein is the remark by Bernard that "it is that which we do know which is the great hindrance to our learning not that which we do not know." The same dilemma faces all creative workers. Byron wrote:

> *To be perfectly original one should think much and read little, and this is impossible, for one must have read before one has learnt to think.*

Shaw's quip "reading rots the mind" is, characteristically, not quite so ridiculous as it appears at first.

The explanation of this phenomenon seems to be as follows. When a mind loaded with a wealth of information contemplates a problem, the relevant information comes to the focal point of thinking, and if that information is sufficient for the particular problem, a solution may be obtained. But if that information is not sufficient—and this is usually so in research—then that mass of information makes it more difficult for the mind to conjure up original ideas, for reasons which will be discussed later. Further, some of that information may be actually false, in which case it presents an even more serious barrier to new and productive ideas.

Thus in subjects in which knowledge is still growing, or where the particular problem is a new one, or a new version of one already solved, all the advantage is with the expert, but where knowledge is no longer growing and the field has been worked out, a revolutionary new approach is required and this is more likely to come from the outsider. The scepticism with which the experts nearly always greet these revolutionary ideas confirms that the available knowledge has been a handicap.

The best way of meeting this dilemma is to read critically, striving to maintain independence of mind and avoid becoming conventionalised. Too much reading is a handicap mainly to people who have the wrong attitude of mind. Freshness of outlook and originality need not suffer greatly if reading is used as a stimulus to thinking and if the scientist is at the same time engaged in active research. In any case, most scientists consider that it is a more serious handicap to investigate a problem in ignorance of what is already known about it.

One of the most common mistakes of the young scientist starting research is that he believes all he reads and does not distinguish between the results of the experiments reported and the author's interpretation of them. Francis Bacon said:

> *Read not to contradict and confute, nor to believe and take for granted...but to weigh and consider.*

The man with the right outlook for research develops a habit of correlating what is read with his knowledge and experience, looking for significant analogies and generalisations. This method of study is one way in which hypotheses are developed, for instance it is how the idea of survival of the fittest in evolution came to Darwin and to Wallace.

Successful scientists have often been people with wide interests. Their originality may have derived from their diverse knowledge. As we shall see in a later chapter on Imagination, originality often consists in linking up ideas whose connection was not previously suspected. Furthermore, variety stimulates freshness of outlook whereas too constant study of a narrow field predisposes to dullness. Therefore reading ought not to be confined to the problem under investigation nor even to one's own field of science, nor, indeed, to science alone. However, outside one's immediate interests, in order to minimise time spent in reading, one can read for the most part superficially, relying on summaries and reviews to keep abreast of major developments. Unless the research worker cultivates wide interests his knowledge may get narrower and narrower and restricted to his own specialty. One of the advantages of teaching is that it obliges the scientist to keep abreast of developments in a wider field than he otherwise would.

It is more important to have a clear understanding of general principles, without, however, thinking of them as fixed laws, than to load the mind with a mass of detailed technical information which can readily be found in reference books or card indexes. For creative thinking it is more important to see the wood than the trees; the student is in danger of being able to see only the trees. The scientist with a mature mind, who has reflected a good deal on scientific matters, has not only had time to accumulate technical details but has acquired enough perspective to see the wood.

Nothing that has been said above ought to be interpreted as depreciating the importance of acquiring a thorough grounding in the fundamental sciences. The value to be derived from superficial and "skim" reading over a wide field depends to a large extent on the reader having a background of knowledge which enables him quickly to assess the new work reported and grasp any significant findings. There is much truth in the saying that in science the mind of the adult can build only as high as the foundations constructed in youth will support.

In reading that does not require close study it is a great help to develop the art of skim-reading. Skimming properly done enables one to cover a large amount of literature with economy of time, and to select those parts which are of special interest. Some styles of writing, of course, lend themselves more to skimming than others, and one should not try to skim closely reasoned or condensed writing or any work which one intends to make the object of a careful study.

Most scientists find it useful to keep a card index with brief abstracts of articles of special interest for their work. Also the preparation of these abstracts helps to impress the salient features of an article in the memory. After reading quickly through the article to get a picture of the whole, one can go back to certain parts, whose full significance is then apparent, re-read these and make notes.

The recent graduate during his first year often studies some further subject in order better to fit himself for research. In the past it has been common for English-speaking research students to study German if they had no knowledge of that language and had already learnt French at school. In the biological sciences I think students would now benefit more from taking a course in biometrics, the importance of which is discussed in the next chapter. In the past it was important to be able to read German, but the output of Germany in the biological and medical sciences has been very small during the last ten years, and it does not seem likely to be considerable for some years to come. Scientists in certain other countries, such as Scandinavia and Japan, who previously often published in the German language, are now publishing almost entirely in English, which, with the vast expansion of science in America as well as throughout the British Commonwealth is becoming the international scientific language. Unless the student of biology has a special reason for wanting to learn German, I think he could employ his time more usefully on other matters until German science is properly revived. In this connection it may be worth noting the somewhat unusual view expressed by the great German chemist, Wilhelm Ostwald, who held that the research student should refrain from learning languages. He considered that the conventional teaching of Latin, in particular, destroys the scientific outlook [1]. Herbert Spencer has also pointed out that the learning of languages tends to increase respect for authority and so discourage development of the faculty of independent judgment, which is so important, especially for scientists. Several famous scientists—including Darwin and Einstein—had a strong distaste for Latin, probably because their independent minds rebelled against developing the habit of accepting authority instead of seeking evidence.

The views expressed in the preceding paragraph on the possible harmful effect of learning languages are by no means widely accepted. However, there is another consideration to be taken into account when deciding whether or not to study a language, or for that matter any other subject. It is that time and effort spent in studying subjects not of great value are lost from the study of some other subject, for the active-minded scientist is constantly faced with what might be called the problem of competing interests: he rarely has enough time to do all that he would like to and should do, and so he has to decide what he can afford to neglect. Bacon aptly said that we must determine the relative value of knowledges. Cajal decries the popular idea that all knowledge is useful; on the contrary, he says, learning unrewarding subjects occupies valuable time if not actual space in the mind [2]. However, I do not wish to imply that subjects should be judged on a purely utilitarian basis. It is regrettable that we scientists can find so little time for general literature.

If the student cannot attend a course in biometrics, he can study one of the more easily understood books or articles on the subject. The most suitable that have come to my notice are those of G.W. Snedecor [3], which deals with the application of statistics to animal and plant experimentation, and A. Bradford Hill [4] which deals mainly with statistics in human medicine. Topley and Wilson's text-book of bacteriology contains a good chapter on the application of biometrics to bacteriology [5]. Professor R.A. Fisher's two books are classical works, but some people find them too difficult for a beginning [6-7]. It is not necessary for the biologist to become an expert at biometrics if he has no liking for the subject, but he ought to know enough about it to avoid either undue neglect or undue respect for it and to know when he should consult a biometrician.

Another matter to which the young scientist might well give attention is the technique and art of writing scientific papers. The general standard of English in scientific papers is not high and few of us are above criticism in this matter. The criticism is not so much against the inelegance of the English as lack of clarity and accuracy. The importance of correct use of language lies not only in being able to report research well; it is with language that we do most of our thinking. There are several good short books and articles on the writing of scientific papers. Trelease [8] deals particularly with the technicalities of writing and editing and Kapp [9] and Allbutt [10] are mainly concerned with the writing of suitable English. Anderson [11] has written a useful paper on the preparation of illustrations and tables for scientific papers. I have found that useful experience can be gained by writing abstracts for publication. Thereby one becomes familiar with the worst faults that arise in reporting scientific work and at the same time one is subjected to a salutary discipline in writing concisely.

The scientist will find his life enriched and his understanding of science deepened by reading the lives and works of some of the great men of science. Inspiration derived from this source has given many young scientists a vision that they have carried throughout their lives. Two excellent recent biographies I can recommend are Dubos' *Louis Pasteur: Freelance of Science* [12] and Marquardt's

Paul Ehrlich [13]. In recent years more and more attention is being given to the study of the history of science and every scientist ought to have at least some knowledge of this subject. It provides an excellent corrective to ever-increasing specialisation and broadens one's outlook and understanding of science. There are books which treat the subject not as a mere chronicle of events but with an insight which gives an appreciation of the growth of knowledge as an evolutionary process (e.g. [14-15]). There is a vast literature dealing with the philosophy of science and the logic of scientific method. Whether one takes up this study depends upon one's personal inclinations, but, generally speaking, it will be of little help in doing research.

It is valuable experience for the young scientist to attend scientific conferences. He can there see how contributions to knowledge are made by building on the work of others, how papers are criticised and on what basis, and learn something of the personalities of scientists working in the same field as himself. It adds considerably to the interest of research to be personally acquainted with the authors of the papers one reads, or even merely to know what they look like. Conferences also provide a good demonstration of the healthy democracy of science and the absence of any authoritarianism, for the most senior members are as Unable to be criticised as is anyone else. Every opportunity should be taken to attend occasional special lectures given by eminent scientists as these can often be a rich source of inspiration. For instance, F.M. Burnet [16] said in 1944 that he had attended a lecture in 1920 by Professor Orme Masson, a man with a real feeling for science, who showed with superb clarity both the coming progress in atomic physics and the intrinsic delight to be found in a new understanding of things. Burnet said that although he had forgotten most of the substance of that lecture, he would never forget the stimulus it conveyed.

Setting about the Problem

In starting research obviously one has first to decide what problem to investigate. While this is a matter on which consultation with an experienced research worker is necessary, if the research student is mainly responsible for choosing his own problem he is more likely to make a success of it. It will be something in which he is interested, he will feel that it is all his own and he will give more thought to it because the responsibility of making a success of it rests on himself. It is wise for him to choose a subject within the field which is being cultivated by the senior scientists in his laboratory. He will then be able to benefit from their guidance and interest and his work will increase his understanding of what they are doing. Nevertheless, if a scientist is obliged to work on a given problem, as may be the case in applied research, very often an aspect of real interest can be found if he gives enough thought to it. It might even be said that most problems are what the worker makes them. The great American bacteriologist Theobald Smith said that he always took up the problem that lay before him, chiefly because of the easy access of material, without which research is crippled [17]. The student with any real talent for research usually has no difficulty in finding a suitable problem. If he has not in the course of his studies noticed gaps in knowledge, or inconsistencies, or has not developed some ideas of his own, it does not augur well for his future as a research worker. It is best for the research student to start with a problem in which there is a good chance of his accomplishing something, and, of course, which is not beyond his technical capabilities. Success is a tremendous stimulus and aid to further progress whereas continued frustration may have the opposite effect.

After a problem has been selected the next procedure is to ascertain what investigations have already been done on it. Textbooks, or better, a recent review article, are often useful as starting points, since they give a balanced summary of present knowledge, and also provide the main references. A textbook, however, is only a compilation of certain facts and hypotheses selected by the author as the most significant at the time of writing, and gaps and discrepancies may have been smoothed out in order to present a coherent picture. One must, therefore, always consult original articles. In each article there are references to other appropriate articles, and trails followed up in this way lay open the whole literature on the subject. Indexing journals are useful in providing a comprehensive coverage of references on any subject to within a year or so of the present, and where they cease a search is necessary in appropriate individual journals. The *Quarterly Cumulative Index Medicus*, *Zoological Record*, *Index Veterinarius* and the *Bibliography of Agriculture* are the standard indexing journals in their respective spheres. Trained librarians know how to survey literature systematically and scientists fortunate enough to be able to call on their services can obtain a complete list of references on any particular subject. It is advisable to make a thorough study of all the relevant literature early in the investigation, for much effort may be wasted if even only one significant article is missed. Also during the course of the investigation, as well as watching for new articles on the problem, it is very useful to read superficially over a wide field keeping constant watch for some new principle or technique that may be made use of.

In research on infectious diseases usually the next step is to collect as much firsthand information as possible about the actual problem as it occurs locally. For instance, if an animal disease is being investigated, a common procedure is to carry out field observations and make personal enquiries from farmers. This is an important prerequisite to any experimental work, and occasionally investigators who have neglected it undertake laboratory work which has little relation to the real problem. Appropriate laboratory examination of specimens is usually carried out as an adjunct to this field work.

Farmers, and probably lay people generally, not infrequently colour their evidence to fit their notions. People whose minds are not disciplined by training often tend to notice and remember events that support their views and forget others. Tactful and searching enquiry is necessary to ascertain exactly what they have observed—to separate their observations from their interpretations. Such patient enquiry is often well repaid, for farmers have great opportunities of gathering information. The important discovery that ferrets are susceptible to canine distemper arose from an assertion of a gamekeeper. His statement was at first not taken seriously by the scientists, but fortunately they later decided to see if there was anything in it. It is said that for two thousand years the peasants of Italy have believed that mosquitoes were concerned with the spread of malaria although it was only about fifty years ago that this fact was established by scientific investigation.

It is helpful at this stage to marshal and correlate all the data, and to try to define the problem. For example, in investigating a disease one should try to define it by deciding what are its manifestations and so distinguish it from other conditions with which it may be confused. Hughlings Jackson is reported to have said:

> *The study of the causes of things must be preceded by the study of things caused.*

To show how necessary this is, there is the classical example of Noguchi isolating a spirochaete from cases of leptospiral jaundice and reporting it as the cause of yellow fever. This understandable mistake delayed yellow fever investigations (but the rumour that it led to Noguchi's suicide has no basis in fact). Less serious instances are not infrequently seen closer at hand.

The investigator is now in a position to break the problem down into several formulated questions and to start on the experimental attack. During the preparatory stage his mind will not have been passively taking in data but looking for gaps in the present knowledge, differences between the reports of different writers, inconsistencies between some observed aspect of the local problem and previous reports, analogies with related problems, and for clues during his field observations. The active minded investigator usually finds plenty of scope for the formulation of hypotheses to explain some of the information obtained. From the hypotheses, certain consequences can usually be proved or disproved by experiment, or by the collection of further observational data. After thoroughly digesting the problem in his mind, the investigator decides on an experiment which is likely to give the most useful information and which is within the limitations of his own technical capacity and the resources at his disposal. Often it is advisable to start on several aspects of the problem at the same time. However, efforts should not be dispersed on too wide a front and as soon as one finds something significant it is best to concentrate on that aspect of the work.

As with most undertakings, the success of an experiment depends largely on the care taken with preliminary preparations. The most effective experimenters are usually those who give much thought to the problem beforehand and resolve it into crucial questions and then give much thought to designing experiments to answer the questions. A crucial experiment is one which gives a result consistent with one hypothesis and inconsistent with another. Hans Zinsser writing of the great French bacteriologist, Charles Nicolle, said:

> *Nicolle was one of those men who achieve their successes by long preliminary thought before an experiment is formulated, rather than by the frantic and often ill-conceived experimental activities that keep lesser men in ant-like agitation. Indeed, I have often thought of ants in observing the quantity output of 'what-of-it' literature from many laboratories...Nicolle did relatively few and simple experiments. But every time he did one, it was the result of long hours of intellectual incubation during which all possible variants had been considered and were allowed for in the final tests. Then he went straight to the point, without wasted motion. That was the method of Pasteur, as it has been of all the really great men of our calling, whose simple, conclusive experiments are a joy to those able to appreciate them.*

Sir Joseph Barcroft, the great Cambridge physiologist, is said to have had the knack of reducing a problem to its simplest elements and then finding an answer by the most direct means. The general subject of planning research is discussed later under the title "Tactics."

Summary

One of the research worker's duties is to follow the scientific literature, but reading needs to be done with a critical, reflective attitude of mind if originality and freshness of outlook are not to be lost. Merely to accumulate information as a sort of capital investment is not sufficient. Scientists tend to work best on problems of their own choice but it is advisable for the beginner to start on a problem which is not too difficult and on which he can get expert guidance. The following is a common sequence in an investigation on a medical or biological problem, (a) The relevant literature is critically reviewed. (b) A thorough collection of field data or equivalent observational enquiry is conducted, and is supplemented if necessary by laboratory examination of specimens. (c) The information obtained is marshalled and correlated and the problem is defined and broken down into specific questions. (d) Intelligent guesses are made to answer the questions, as many hypotheses as possible being considered, (e) Experiments are devised to test first the likeliest hypotheses bearing on the most crucial questions.

References

[1] Ostwald, W. *Die Forderung der Tages.* (Leipzig: Verlagsellschaft mbH, 1910).
[2] Cajal, S. Ramón y *Precepts and Counsels on Scientific Investigation, Stimulants of the Spirit.* Trans by J.M. Sanchez-Perez. (Mountain View, CA: Pacific Press Publ. Assn., 1951)
[3] Snedecor, G.W. *Statistical Methods applied to Experiments in Agriculture and Biology.* (Ames, IA: Collegiate Press Inc., 1938).
[4] Bradford Hill, A. *The Principles of Medical Statistics.* (London: The Lancet Ltd., 1948).
[5] Topley, W.W. C, & Wilson, G.S. *The Principles of Bacteriology and Immunity.* (London: Edward Arnold & Co., 1929).
[6] Fisher, R.A. *The Design of Experiments.* (London: Oliver & Boyd, 1935).
[7] Fisher, R.A. *Statistical Methods for Research Workers.* (London: Oliver & Boyd, 1938).
[8] Trelease, S.F. *The Scientific Paper; How to Prepare it; How to Write it.* (Baltimore: Williams & Wilkins Co., 1947).
[9] Kapp, R.O. *The Presentation of Technical Information.* (London: Constable & Co., 1948).
[10] Allbutt, C.T. *Notes on the Composition of Scientific Papers.* (London: Macmillan & Co. Ltd., 1905).
[11] Anderson, J.A. "The preparation of illustrations and tables." *Trans. Amer. Assoc. Cereal Chem.*, No. 3, 74 (1945).
[12] Dubos, Rene J. *Louis Pasteur: Free Lance of Science.* (Boston: Little, Brown & Co., 1950).
[13] Marquardt, M. *Paul Ehrlich.* (London: Wm. Heinemann Ltd., 1949)

[14] Butterfield, H. *The Origins of Modern Science, 1300-1800.* (London: G. Bell & Sons Ltd., 1949).
[15] Conant, J.B. *On Understanding Science. An Historical Approach.* (London: Oxford Univ. Press, 1947).
[16] Burnet, F.M. *Bulletin of the Australian Association of Scientific Workers*, No. 55 (1944)
[17] Smith, T. *Journal of Bacteriology* Vol. 27, 19 (1934).

3

Ten Simple Rules for Getting Published
Philip E. Bourne

Philip E. Bourne is Editor-in-Chief of PLoS Computational Biology and is a professor in the Skaggs School of Pharmacy and Pharmaceutical Sciences at the University of California San Diego and Associate Vice Chancellor for Innovation and Industrial Alliances. He is also Associate Director of the RCSB Protein Data Bank (PDB) and is well-known for his work in advancing techniques of computational bioinformatics. In this article in PLoS Computational Biology, he gives advice on how young scientists can make sure they get published. This is part of the "Ten Simple Rules" series.

The student council (http://www.iscbsc.org/) of the International Society for Computational Biology asked me to present my thoughts on getting published in the field of computational biology at the Intelligent Systems in Molecular Biology conference held in Detroit in late June of 2005. Close to 200 bright young souls (and a few not so young) crammed into a small room for what proved to be a wonderful interchange among a group of whom approximately one-half had yet to publish their first paper. The advice I gave that day I have modified and present as ten rules for getting published.

Rule 1: Read many papers, and learn from both the good and the bad work of others.

It is never too early to become a critic. Journal clubs, where you critique a paper as a group, are excellent for having this kind of dialogue. Reading at least two papers a day in detail (not just in

---

* Reprinted from P.E. Bourne, Ten Simple Rules for Getting Published. *PLoS Computational Biology* Vol. 1 No. 5: e57 (2005); Creative Commons Attribution License

your area of research) and thinking about their quality will also help. Being well read has another potential major benefit—it facilitates a more objective view of one's own work. It is too easy after many late nights spent in front of a computer screen and/or laboratory bench to convince yourself that your work is the best invention since sliced bread. More than likely it is not, and your mentor is prone to falling into the same trap, hence rule 2.

Rule 2: The more objective you can be about your work, the better that work will ultimately become.

Alas, some scientists will never be objective about their own work, and will never make the best scientists—learn objectivity early, the editors and reviewers have.

Rule 3: Good editors and reviewers will be objective about your work.

The quality of the editorial board is an early indicator of the review process. Look at the masthead of the journal in which you plan to publish. Outstanding editors demand and get outstanding reviews. Put your energy into improving the quality of the manuscript *before submission*. Ideally, the reviews will improve your paper. But they will not get to imparting that advice if there are fundamental flaws.

Rule 4: If you do not write well in the English language, take lessons early; it will be invaluable later.

This is not just about grammar, but more importantly comprehension. The best papers are those in which complex ideas are expressed in a way that those who are less than immersed in the field can understand. Have you noticed that the most renowned scientists often give the most logical and simply stated yet stimulating lectures? This extends to their written work as well. Note that writing clearly is valuable, even if your ultimate career does not hinge on producing good scientific papers in English language journals. Submitted papers that are not clearly written in good English, unless the science is truly outstanding, are often rejected or at best slow to publish since they require extensive copyediting.

Rule 5: Learn to live with rejection.

A failure to be objective can make rejection harder to take, and you will be rejected. Scientific careers are full of rejection, even for the best scientists. The correct response to a paper being rejected or requiring major revision is to listen to the reviewers and respond in an objective, not subjective, manner. Reviews reflect how your paper is being judged—learn to live with it. If reviewers are unanimous about the poor quality of the paper, move on—in virtually all cases, they are right. If they request a major revision, do it and address every point they raise both in your cover letter and through obvious revisions to the text. Multiple rounds of revision are painful for all those concerned and slow the publishing process.

Rule 6: The ingredients of good science are obvious—novelty of research topic, comprehensive coverage of the relevant literature, good data, good analysis including strong statistical support, and a thought-provoking discussion. The ingredients of good science reporting are obvious—good organization, the appropriate use of tables and figures, the right length, writing to the intended audience—do not ignore the obvious.

Be objective about these ingredients when you review the first draft, and do not rely on your mentor. Get a candid opinion by having the paper read by colleagues without a vested interest in the work, including those not directly involved in the topic area.

Rule 7: Start writing the paper the day you have the idea of what questions to pursue.

Some would argue that this places too much emphasis on publishing, but it could also be argued that it helps define scope and facilitates hypothesis-driven science. The temptation of novice authors is to try to include everything they know in a paper. Your thesis is/was your kitchen sink. Your papers should be concise, and impart as much information as possible in the least number of words. Be familiar with the guide to authors and follow it, the editors and reviewers do. Maintain a good bibliographic database as you go, and read the papers in it.

Rule 8: Become a reviewer early in your career.

Reviewing other papers will help you write better papers. To start, work with your mentors; have them give you papers they are reviewing and do the first cut at the review (most mentors will be happy to do this). Then, go through the final review that gets sent in by your mentor, and where allowed, as is true of this journal, look at the reviews others have written. This will provide an important perspective on the quality of your reviews and, hopefully, allow you to see your own work in a more objective way. You will also come to understand the review process and the quality of reviews, which is an important ingredient in deciding where to send your paper.

Rule 9: Decide early on where to try to publish your paper.

This will define the form and level of detail and assumed novelty of the work you are doing. Many journals have a presubmission enquiry system available—use it. Even before the paper is written, get a sense of the novelty of the work, and whether a specific journal will be interested.

Rule 10: Quality is everything.

It is better to publish one paper in a quality journal than multiple papers in lesser journals. Increasingly, it is harder to hide the impact of your papers; tools like Google Scholar and the ISI Web of Science are being used by tenure committees and employers to define metrics for the quality of your work. It used to be that just the journal name was used as a metric. In the digital world, everyone knows if a paper has little impact. Try to publish in journals that have high impact factors; chances are your paper will have high impact, too, if accepted.

When you are long gone, your scientific legacy is, in large part, the literature you left behind and the impact it represents. I hope these ten simple rules can help you leave behind something future generations of scientists will admire.

4

Three Essays from *What is Life*
J.B.S. Haldane

John Burdon Sanderson (J.B.S.) Haldane (November 5, 1892 – December 1, 1964)—was a renowned mathematical biologist and along with R.A. Fisher and Sewall Wright is considered one of the founders of population genetics and the Modern Synthesis, which unified Darwin's Theory of Evolution by Natural Selection and Mendelian genetics. While a renowned scientist, he was also often controversial due to his Marxist beliefs and was criticized for his tepid criticism of the ideas of Lysenko in the Soviet Union. In his book, "What Is Life?", he discusses many valuable ideas, minus political comments, on how to do science and what sort of training best suits scientists.

Essay 1: What Use is Science to You?

I have just been reading Prof. G.H. Hardy's *A Mathematician's Apology*, in which he states the case for pure mathematics as a life work. He regards himself as essentially an artist, creating a peculiar kind of beauty, as a musical composer does. He believes that his work has had no practical value. This is probably true today. But experience suggests that fifty or a hundred years hence it may be of great practical use, when much of our existing applied mathematics is as out of date as that used by Drake for navigation. He then makes the remarkable statement that science in general is of no use except for specialists.

> *It is useful*, he writes, *to be tolerably quick at arithmetic. It is useful to know a little French and German, a little history and geography, perhaps even a little economics. But a little chemistry, physics, or physiology has no value at all in ordinary life. We know that the gas will burn without knowing its constitution; when our cars break down we take them to a garage; when our stomach is out of order; we go to a doctor or a drug store.*

---

* Reprinted from J.B.S. Haldane *What is Life?* (New York: Boni and Gaer, 1947) 119-122; 134-137; 145-148

I am inclined to think Prof. Hardy's statement is largely true. But it is not so much a statement about science, as about our existing society. Professor Hardy is a university don, and an extreme specialist. Dons are encouraged to specialize, and to fill their spare time with some harmless recreation, such as the collection of Japanese prints, the cultivation of roses, or a scholarly knowledge of cricket scores of the past. Workers are also encouraged to be specialists, skilled at a particular process, but not understanding the principles involved in it, still less how their factory as a whole is run. Only a few progressive firms encourage their workers to get a scientific education. Lenin aimed at something quite different, a society of "men who can do everything," where professors can and do mend their own cars, and mechanics investigate the scientific principles involved in their jobs. The Soviet Union has not yet reached this goal, but it is well on the way to it. Foreign engineers who brought over complicated machinery to the Soviet Union were often horrified because the workers insisted on taking it to bits, even if it took a month to put it together again. This meant a delay in starting production, but it meant that the workers were able to mend the machines if they went wrong, and often to improve on their design fairly quickly. Soviet intellectuals are mostly drawn from the ranks of the manual workers, and proud to show that they are still capable of tackling skilled jobs. Clearly a society of this kind can switch over from peace to war, and will switch over from war to peace again, much more quickly than our own; quite apart from the fact that there are no landlords and other vested interests to impede these changes.

In that kind of society, if not in our own, a moderate acquaintance with science will be of practical use to everybody. But scientific education also has a moral value. A child studies English literature and learns what to write about Wordsworth in order to get a school certificate. He or she may think that Wordsworth was an intolerable old bore and humbug, but if his or her future depends on a certificate, this view must be kept dark. French is about as bad. I can make a public speech in French (or near enough to French for Frenchmen to follow it) and read any French book, but I doubt if I could pass a school certificate in

French, because (like a good many Frenchmen) I have forgotten some of the rules of grammar.

But in science one can always go behind the teacher to nature. Here is a lens whose focal length can be measured, a rabbit whose ribs can be counted, or a dynamo which will or will not work. Nature does not accept excuses, as you find when you put in too much sulphuric acid, and it splashes over your face. History does not do so either, though historians do their best to justify "good men but bad kings." A scientific education, provided it is a real education with practical work in which the pupil gets a chance of making mistakes, rubs it into you that good intentions are not enough; good technique is needed as well. This is why Marxism is rightly called scientific socialism. It is based not merely on an analysis of the breakdown of capitalism, and what is needed to replace it, but on a study of how social changes actually do occur. Before 1917 it had not had an adequate experimental test. But Lenin, Stalin, and their collaborators, have given it the test which every scientific theory requires, and it has proved to work in practice. No wonder that Marxism is spreading among scientists to an extent which alarms many people. No wonder also that reactionaries insist on the moral value classical education, that is to say, an education in which the test of correctness is tradition, not experiment. They also do their best to make scientific education; especially in schools, fit into the same mold, and some teachers succeed in making botany as dull as Latin grammar. Scientific education is largely useless unless it teaches one to make one's own experiments and observations and to trust in their results, and unless it is linked up with the actual practice of industry, agriculture, or medicine. If this is achieved, there is plenty of room for the literary and historic side of education too.

But the products of such an education will form their own opinions on history and literature. And they are a good deal more likely to change the course of history, and to write great literature themselves, than those who have absorbed an education designed to fit them for a social system which is falling to pieces around them.

Essay 2: The Back Door to Science

Most of the men and women who are engaged in scientific research get university degrees in science. It is sometimes thought that this is necessary. This is not the case. I only got a scientific degree this year, an honorary one from the Dutch university of Groningen. However, I had an Arts Degree at Oxford, and it is certainly hard to get a paid research post without a degree of some kind. In consequence thousands of men and women who could do good research work are doing other things because they could not afford a university education. Scholarships are better than nothing, but they certainly do not pick out the best future research workers. Even university honors examinations do not do so. A good many people who get first classes show no originality, and others who get lower classes do highly original research. Scholarships select children for precocity. The child who gets into a secondary school as the result of an examination at the age of eleven, may have grown up mentally quicker than the average. It does not follow that she will go on doing so. They also select for home environments where intellectual work is fairly easy. An only child has much better chances of doing home work than one with half a dozen brothers and sisters, but that does not make her a better research worker.

Now there is an opening into scientific research for those who have been unable to get scholarships, but feel the pull of science. This is by becoming a laboratory technician in the right kind of laboratory. In many laboratories there is no great future for technicians. At best they may hope to become head technicians supervising routine work, or perhaps constructing apparatus with which others will carry out research. At worst they will find themselves in poorly paid posts doing semi-skilled work. This was the case until fairly recently, in most university laboratories. The change which is now taking place is very largely due to the Association of Scientific Workers, which includes technicians as well as university graduates, and has fought their battles in many places. At University College, London, for example, our junior technicians not merely have the right to education, but are compelled to put in at least eight hours a week in various classes. I think that some of our young ladies who hope to marry as soon as possible resent this considerably. These classes include lectures and practical work. We hope to give the technicians a thorough grounding in several branches of science and in mathematics. In our chemistry department a substantial fraction of the technicians ultimately get degrees. It takes them a good deal longer than the ordinary student to do so. And I have little doubt that in the examinations their practical work is better than the average, and their written work worse.

A few of the more ambitious technicians go to evening classes at a Technical College or at Birkbeck College. This year one of them got the best first class of his year in London University in his subject, and was immediately appointed as a University Lecturer. This of course leads to us losing many of our best technicians. In the Soviet Union, where technicians have, I am told, a higher status than in University College, such loss is a real handicap to senior research workers. However, even from a selfish point of view, it will probably be worth our while to get ambitious and intelligent young men and women, who will reach a very high standard in their work. Meanwhile thousands of boys and girls who want to take up science and would normally be able to join a university as students are kept out by the flood of ex-service men and women, who are quite rightly given priority. If any of them read this, I recommend them to think seriously about taking a post as a technician. Of course there are still many institutions where the technicians have no chance. But there are others where they have. The Association of Scientific Workers should be able to tell them what to expect in any particular case.

I recommend it for the following reasons. I learned much of my practical science by "bottle washing" for my father. Washing laboratory glassware is a highly skilled job. It is one thing to make glassware clean enough for ordinary chemical analysis, and quite another to prepare it for bacteriological work, where a single bacillus left behind may falsify an experiment. I went on to make standard solutions calibrate apparatus, and so on, reaching a far higher standard of accuracy than is demanded in a teaching laboratory or in many industrial ones. In fact I learned a good deal of science as an apprentice rather than as a student. Scientific research demands manual skill as well as thought. So it is very important that a number of the recruits to it should be primarily qualified by their skill, even if they find the theoretical side of science difficult or dull.

There are no vacancies for students at present, but a fair number for technicians. Some of our best technicians will be satisfied with turning out skilled work. It is very pleasant to make really good microscope slides, to grow really good crystals, and above all to make really accurate apparatus. Naturally a professor is delighted to get a man or woman with no further ambitions. But we also realize that many of our helpers will want to discover something for themselves and that it is our duty both to the technicians and to science to help them on their path. I cannot set up an employment bureau. Above all I don't know whether Professor Smith gives technicians a chance, while Professor Jones does not. I only know that among my colleagues there are men who have started as technicians, not as university students and have made good. I write this article because I want to see more of them.

Essay 3: Can Science be Taught?

One of my jobs in peace time is to lecture and hold practical classes on certain branches of biology. This is not essentially different from the task of a secondary school teacher. In each case our pupils are working for examinations, even though many of them are sufficiently interested to study some matters on which they will not be examined. Are we really doing the best we could to turn out the scientists of the next generation? By scientists I mean men and women who actually advance science, either by discovering new facts about nature or by applying those which were known before to practical problems. Of course the distinction is not a sharp one.

I am inclined to say "No," if only because I have had a fairly successful scientific career without passing any examinations in science since I left school. I had a vastly better training than any university course, namely apprenticeship. The teaching of science involves a contradiction. The very core of the scientific attitude is a respect for facts, whether or not they agree with the teaching of the Bible, of Darwin, or of Engels. But in school one is taught on the authority of the teacher or the textbook. A real teacher of science must persuade his pupils that his lectures may be incorrect, and are bound to be so at some points; whereas the spelling of English or French can and should be taught authoritatively. I learned my most important lessons in science from my father. When I was about eight, I began taking down figures which he dictated to me during analyses. By the time I was twelve I was taking samples of air in mines, and mixing soda and lime for rescue apparatus. Later I came on to the really responsible job of bottle-washing. I realized that the standards of accuracy required in real research, where a mistake might lose a life, were vastly higher than those of school chemistry, just as my own laboratory standard of cleanliness was above that of the scullery. I heard my father producing beautiful theories which he had to scrap in deference to ugly facts. I got at least a rough idea of what scientific research meant.

The most important part of the science teaching which I have done was probably the supervision of young workers doing their first research. They must be allowed to make mistakes—even bad ones which will waste a month—but yet helped, if necessary, so that they get some results from their work. Some of my junior colleagues have done pretty well, and at least two are probably better biologists than myself.

If we are to teach children science, as we do teach them English, up to the point where some of them show at least a little originality, we must introduce something of this kind into' schools. This has already been done in many Soviet schools, though presumably their scientific teaching has suffered from the war. In some country schools a party of boys and girls would accompany a geologist on a week's prospecting tour, finding samples of minerals and fossils. In others they helped in the collection of animals and plants. Many schools cooperated in a study of bird migration. Thousands of birds were caught in traps specially designed not to hurt them, released with rings on their legs, and caught again at other schools. When all the results were collated, it was possible to map the routes taken across the Union by various species of birds. In many of the towns workshops were available where children could test their own inventions. Probably very few of these were much use, but at least they learned from their failures. And they had the satisfaction of making something from their own design, instead of merely repeating standard exercises.

How far could our secondary school children learn science in this way after the war? The first requisite is a much bigger and better supply of science teachers, particularly men and women who have done a little research, if only up to the M.Sc. standard. One of the main tasks of our universities after the war should be to produce such teachers. The second is more leisure for the teachers, and for the brightest ten per cent of boys and girls learning science. This would be available if free university education were provided on the scale of the Soviet Union, or even the United States. If so the ablest children would have qualified for it by the age of 15, and would have some time to spare in their last school years.

Under socialism such things would be vastly easier. The state-owned factories and railways could and would be linked up with the schools in a way which is impossible at present. School children would be more welcome on an estate belonging to their own parents than in Lord Blank's pheasant preserves or Sir John Dash's Home Farm.

Many teachers may think that, at best, such work will not make a real contribution to knowledge. If they think so it is because they themselves have been taught science badly, and do not realize how little we know about quite simple matters. Professor Salisbury, who has just been appointed director of Kew Gardens, has recently published the first comprehensive counts of the numbers of seeds produced by common English plant species. Such figures are most important for the theory of evolution, but had never been compiled before. On an average only one seed per year from an annual plant can germinate and grow up into a new plant. But this one is one in 70,000 with the field poppy, and about one in 30 with the cuckoo-pint. Clearly natural selection is much more intense in the poppy than the cuckoo-pint, even though the latter lives several years. Children could easily undertake such work as this.

Under the new educational schemes children are to be segregated into grammar, modem, and technical schools at the age of eleven. Presumably in the modern schools they will be taught science by the existing methods, while in the technical schools they will have a chance of learning the craftsmanship which every scientist needs, but few scientific principles. In fact, the gap between theory and practice, which is characteristic of capitalism or any other class society, will be made a feature of our education. This may help to delay the coming of a classless society. But it will not give us the scientists whom we shall need if we are to hold our place among the nations.

5

Four Golden Lessons
Steven Weinberg

Steven Weinberg is a theoretical physicist at the University of Texas Austin. One of the great minds of 20th century physics, he along with Abdus Salam and Sheldon Glashow unified the electromagnetic and weak nuclear forces into the electroweak force. For this he received the 1979 Nobel Prize in Physics and helped renew hopes amongst physicists that all four fundamental forces—gravity, electromagnetism, the weak nuclear force, and the strong nuclear force could be unified into a comprehensive description of the laws of physics. A longtime passionate advocate of science education, in this article, he gives advice to young scientists on how to approach the process of discovery.

When I received my undergraduate degree—about a hundred years ago—the physics literature seemed to me a vast, unexplored ocean, every part of which I had to chart before beginning any research of my own. How could I do anything without knowing everything that had already been done? Fortunately, in my first year of graduate school, I had the good luck to fall into the hands of senior physicists who insisted, over my anxious objections, that I must start doing research, and pick up what I needed to know as I went along. It was sink or swim. To my surprise, I found that this works. I managed to get a quick PhD—though when I got it I knew almost nothing about physics. But I did learn one big thing: that no one knows everything, and you don't have to.

Another lesson to be learned, to continue using my oceanographic metaphor, is that while you are swimming and not sinking you should aim for rough water. When I was teaching at the Massachusetts Institute of Technology in the late 1960s, a student told me that he wanted to go into general relativity rather than the area

*Reprinted with permission of the Nature Publishing Group from Steven Weinberg,"Scientist: Four golden lessons" *Nature* 426, 389 (27 November 2003)

I was working on, elementary particle physics, because the principles of the former were well known, while the latter seemed like a mess to him. It struck me that he had just given a perfectly good reason for doing the opposite. Particle physics was an area where creative work could still be done. It really was a mess in the 1960s, but since that time the work of many theoretical and experimental physicists has been able to sort it out, and put everything (well, almost everything) together in a beautiful theory known as the standard model. My advice is to go for the messes—that's where the action is.

My third piece of advice is probably the hardest to take. It is to forgive yourself for wasting time. Students are only asked to solve problems that their professors (unless unusually cruel) know to be solvable. In addition, it doesn't matter if the problems are scientifically important—they have to be solved to pass the course. But in the real world, it's very hard to know which problems are important, and you never know whether at a given moment in history a problem is solvable. At the beginning of the twentieth century, several leading physicists, including Lorentz and Abraham, were trying to work out a theory of the electron. This was partly in order to understand why all attempts to detect effects of Earth's motion through the ether had failed. We now know that they were working on the wrong problem. At that time, no one could have developed a successful theory of the electron, because quantum mechanics had not yet been discovered. It took the genius of Albert Einstein in 1905 to realize that the right problem on which to work was the effect of motion on measurements of space and time. This led him to the special theory of relativity. As you will never be sure which are the right problems to work on, most of the time that you spend in the laboratory or at your desk will be wasted. If you want to be creative, then you will have to get used to spending most of your time not being creative, to being becalmed on the ocean of scientific knowledge.

Finally, learn something about the history of science, or at a minimum the history of your own branch of science. The least important reason for this is that the history may actually be of some use to you in your own scientific work. For instance, now and then scientists are hampered by believing one of the over-simplified models of science that have been proposed by philosophers from Francis Bacon to Thomas Kuhn and Karl Popper. The best antidote to the philosophy of science is a knowledge of the history of science.

More importantly, the history of science can make your work seem more worthwhile to you. As a scientist, you're probably not going to get rich. Your friends and relatives probably won't understand what you're doing. And if you work in a field like elementary particle physics, you won't even have the satisfaction of doing something that is immediately useful. But you can get great satisfaction by recognizing that your work in science is a part of history.

Look back 100 years, to 1903. How important is it now who was Prime Minister of Great Britain in 1903, or President of the United States? What stands out as really important is that at McGill University, Ernest Rutherford and Frederick Soddy were working out the nature of radioactivity. This work (of course!) had practical applications, but much more important were its cultural implications. The understanding of radioactivity allowed physicists to explain how the Sun and Earth's cores could still be hot after millions of years. In this way, it removed the last scientific objection to what many geologists and paleontologists thought was the great age of the Earth and the Sun. After this, Christians and Jews either had to give up belief in the literal truth of the Bible or resign themselves to intellectual irrelevance. This was just one step in a sequence of steps from Galileo through Newton and Darwin to the present that, time after time, has weakened the hold of religious dogmatism. Reading any newspaper nowadays is enough to show you that this work is not yet complete. But it is civilizing work, of which scientists are able to feel proud.

PART II
Creative Thinking & Reason

6

Creativity and Intuition Excerpt from Chapter 3: The Conception and Experience of Creativity Hideki Yukawa

Hideki Yukawa (January 23, 1907 – September 8, 1981)—was a Japanese theoretical particle physicist and Japan's first Nobel laureate. Yukawa came of age in Kyoto during a tumultuous time in Japan from the ultimate failure of the Taisho democracy to the militarist regime and World War II. In 1934, at the age of 27, he wrote a paper predicting the existence of a new particle that would mediate the strong nuclear force, the force which keeps protons bound in the nucleus despite having the same charge. He called this particle the meson and its discovery in 1947 earned him the Nobel Prize in 1949. In 1973, he wrote a book, Creativity and Intuition: A Physicist Looks at East and West, a compilation of essays to discuss his ideas on the creative process in science. An excerpt from chapter 3 is included here.

The Breaking Down of Fixed Ideas

This means that there is also something that needs breaking down within ourselves. The need for reform is not without, but within. Einstein's discovery of the principle of relativity was unquestionably a major manifestation of creativity, but this does not mean that Einstein was possessed of the theory of relativity from the time he was born. It was contact with negative experimental results contradicting the ether hypothesis, among other things, that made him change his own way of thinking. This means, to put it differently, that prior to creativity there has to be a struggle with one's own self.

---

* Reprinted with permission of copyright holder Harumi Yukawa from Hideki Yukawa *Creativity and Intuition: A Physicist Looks at East and West* (New York: Kondansha, Inc., 1978) 124-142. Translated by John Bester

A considerable period of preparation is necessary before a particular man can display creativity in a particular field and in a particular form. He must, in short, have acquired all kinds of knowledge and also, probably, have undergone all kinds of training. It is only after many kinds of prior conditions have been satisfied that creativity can show itself. By the time one has done research for a long continuous period and become a full-fledged research worker, one has developed within oneself a relatively stable system of knowledge. This system of knowledge has been integrated by one's own efforts into a particular, definite form. And this business of integrating by oneself is, of course, an extremely valuable experience in itself. It means that one is able to teach others, and to pass on one's own knowledge.

That same state of affairs also means, conversely, that one has become set in one's way of thinking. To exaggerate a little, one has become a mass of fixed ideas. Anyone who carries on his studies for a long enough period of years—myself included—becomes such a mass of fixed ideas.

To know a lot of things has the advantage that, in theory at least, it serves as a basis for discovering new things; but it also has a gradual immobilizing effect. Whatever happens, nothing surprises one; and the chances for a display of creativity are lost.

Drawing Out Latent Ability

In a very large number of cases, creativity is judged by its results. Einstein, for example, discovered the principle of relativity. Even those who do not know what it is all about are convinced that Einstein was an exceptional genius and that an exceptional display of creativity took place.

When one reads in his biography that he showed no particular brilliance when he was young, or that he failed in some examination or other, one's admiration for Einstein increases still further. One would be much less impressed if he had always been at the top of his class. It is much more flattering to many people's egos to think that someone who subsequently made a great discovery had flunked his exams at least once. But if he had not made his great discovery, if he had not become a famous scientist, then the verdict, to the end, would have been that he had never cut much of a figure right from his schooldays.

To judge by results is perhaps only natural, but this will not help to elucidate the essential nature of creativity. One must consider, rather, why there was such a manifestation of creativity, what the state of affairs was before its manifestation, and where it had lain hidden all the while.

It is not something that appears out of the blue. Heredity, environment, and the like doubtless all play their part, but however one cares to express it, the most important thing is that there has always existed the possibility of such creativity appearing, that something hidden, something latent, should appear, should become manifest. Thus the question of creativity, I feel, can ultimately be reduced to the question of where creativity lies hidden, and of the means whereby it can be brought out into the open.

Geniuses Appear in Batches

The seventeenth century saw the emergence of a large number of geniuses. Within the space of one hundred years, an extraordinary number of geniuses—surpassing geniuses, one might say—appeared, beginning with men such as Bacon, Galileo, Kepler, and Descartes and carrying on to Newton and Leibniz.

The beginning of the twentieth century is another case in point, since it produced within a short space of time men such as Planck, Einstein, Rutherford, de Broglie, Born, Heisenberg, Bohr, Schrödinger and Dirac. It is usual, it seems, for geniuses to appear in batches; on the other hand, there are also periods during which the appearance of geniuses is very rare. There must be some reason for this, I feel, other than coincidence.

To take a more everyday example of a similar kind of thing, it is frequently observable in schools that a certain grade or thereabouts will produce a sudden burst of comparatively outstanding young people, which will be followed by a period in which there are none, to be followed again after a while by another similar burst. There are various reasons for this phenomenon, I imagine, but one that can be grasped very easily is the psychological effect. The presence of classmates who work hard and excel in their studies is an encouragement or stimulation to emulate them. This kind of influence probably plays quite a large part.

In the same way it seems likely that on a large scale—over a period of several years, or even of a century—scholars should have a great effect on each other and produce a steady succession of great minds.

The Situation Today

The present age is marked by an extraordinarily large number of scientists. The number of those alive at present probably far exceeds the number of those who have lived and died during the period from ancient times to the present. If the chances of a given individual manifesting creativity were governed by the same laws of probability as the casting of a dice the present age should have produced far more great discoveries and great inventions than it in fact has. Why, then, does the reality seem to be otherwise?

I should like here to touch on the situation in Japan, where, if we go back to ancient times, we notice that the higher manifestations of culture were not indigenous but had been brought in from outside.

From two to two and a half millennia ago, for example, Japan's neighbor China produced a large number of great thinkers; there were many systems of thought that still hold interest for us today, and many philosophers whom we may still call great. Japan at that time was scarcely at a stage to produce any organized philosophy. The first appearance of anything that one might call indigenous and systematic thought in Japan occurred with Kōbō Daishi in the ninth century.

This does not mean any intellectual failing on the part of the Japanese, but simply that in ancient times the time lag between the advanced and backward areas of the world was extremely great. It was because this state of affairs was brought painfully home to the Japanese around the end of the Edo period that they decided to take over Western civilization as speedily as possible.

The influence of such a history on the outlook of the Japanese has in some ways been very persistent. The idea took hold that there was little point in doing creative work at the highest level; that it was better to let foreigners take the lead, and even nowadays reliance on foreign inventions still seems to be too great. Historically and statistically this is correct. Yet nothing profitable is going to result so long as one deals with the question of creativity in terms of statistics alone.

The basic essential here, I feel, is to break down this set way of thinking. This is surely relevant to the display of creativity. Such things, it is true, just cannot be broken down overnight, but I try to persuade myself that the situation is beginning to change a little, and I go on stressing the need without getting too pessimistic.

No one can say whether a certain person originally had creativity or not. As I have already said, we judge by the results, which are extremely difficult to predict.

It seems unlikely that modern man is inferior to ancient man. Modern man has many advantages over his predecessor. If, despite this, few people display creativity, it means that the mental attitudes of those of us who are alive today are wrong. Mental attitudes, if not everything, are at least one important element—I myself would say the most important of all.

Failure as a Source of Creativity

Even to the best-known scholar, success does not come so many times a lifetime. If one does research for, say forty years, one is likely to achieve real success only two or three times in all. Some men go a whole lifetime without achieving a single major success. What were they doing all the time then? Not nothing, of course.

In long years of research, the right opportunity comes along very seldom. Even when it comes, one often fails to grasp it. The result in short is that one seldom succeeds and so the manifestations of creativity are extraordinarily valuable. If success was easily gained, there would be little need to talk of creativity at all. Even the man who, seen from the outside, seems to have had a reasonable amount of success has almost certainly experienced repeated failure. The failures, of course are in no sense wasted. One's failures provide a later source of success. Nor need it only be one's own failures; sometimes the failures of others prove most significant. What others have tried and given up one alters a little and one succeeds. As the old saying goes, "Failure is the mother of success."

I myself often work hard from morning till dusk only to throw what I have done into the wastepaper basket in despair. Quantitatively, the amount I keep because something might come of it is incomparably smaller than what I throw away, yet it is this, I believe, that serves as the basis for creation.

Tenacity and Thinking within a Framework

I have suggested above that the reason why so little creativity is displayed despite the large number of scholars and the advantages they enjoy over their predecessors is that there is something wrong in the mental attitudes of the Japanese. Another reason—though this is susceptible to change—is the fact that the environment in which we find ourselves is extremely unfavorable to concentrating one's efforts for long on the same thing.

Looking back over my own life, for example, I find that the number of miscellaneous tasks I have had to perform has gradually increased over the years. Wherever possible I try nowadays not to take on tasks that fall under this heading, but still I have too many.

On top of this, there is too much information. The bewildering influx of new stimuli leaves one no room to think about things in a relaxed way. Sometimes one tends—is obliged to—flit from one thing to another. What I find particularly awkward is that this information comes to one already arranged and standardized.

When the amount of information coming in becomes excessive, it becomes quite impossible to present it just as it stands. It would be an enormous labor for the individual to set it in order by himself. In practice, and for better or for worse—for worse, in many cases—it has already been set in order for us. The news in the papers and on the television, for example, is a perfect example of this process at work. For convenience's sake, we accept such things in the form in which they are served to us. To do so gradually becomes a habit; it seems to make life easier, yet it also steadily increases one's dependence on others.

At the same time the very fact of arranging, whoever has been responsible, implies the existence of some method or framework.

Even when a person does the arrangement himself, so long as he thinks of things within a fixed framework there is no creativity.

All major creation begins with getting outside the framework, or changing the framework itself.

The important thing if one is to achieve creativity, I feel, is to keep plugging away at one thing despite all the miscellaneous tasks and the superfluity of information with which everyday life seeks to claim one's attention. What is needed, in other words, is tenacity of purpose.

Physicists who show creativity usually work with extraordinary—almost unnecessary, one might think—tenacity at a particular subject. This might seem to contradict what I have just said concerning the need to break down fixed concepts, and it is an extremely tricky point, yet the fact is that most of them have some ideal, image, or vision, which they cling to with unusual pertinacity.

I, for one, have my own particular subject; in the years since I first took up physics, there have been many changes and much development, yet essentially I have been concerned with the same thing throughout.

The question is, to what one should attach oneself in this way? Subjectively speaking, anything should do equally well, from pinball games to scholarship. Yet objectively viewed, it is more meaningful to devote oneself to something as difficult of attainment as possible. One may suffer one's whole life as a result, but it is well worthwhile.

One further question is that the act of devoting one's tenacity to one thing also means, in itself, that one is aware of some contradiction within oneself. Without some contradiction within, there can be no study; that, indeed, is the essential nature of study. To put it differently, one has some place that is dark, or obscure, or vague, or puzzling within oneself, and one tries to find some light in it. Then, when one has found a ray of light, one tries to enlarge it little by little so that the darkness is gradually dispelled.

This, I feel, is the typical process whereby creativity shows itself.

Custom, Imitation, and Creativity

Around 1925, the year preceding that in which he did such important work on quantum mechanics, the great physicist Schrödinger wrote an article concerning his view of the world. In this article, he writes of the consciousness as a kind of outpost of the unconscious, a light that flares up suddenly in the darkness, in the night of the unconscious. This relates directly to the question of creativity, for consciousness is generally connected with some new or fresh experience.

If one goes on repeating the same thing, it gradually becomes unconscious. One comes to do it by reflex, or perhaps one should say by habit. It is the escape from habit of this kind that constitutes creativity. All kinds of habits are firmly established within us, forming a kind of fixed system, and when the consciousness comes into contact with something that does not fall within that system of habit, it often becomes extremely acute.

Another thing is that men copy each other. It is unconscious copying of those about us that helps us grow into adults. We are obliged to copy in all kinds of ways, otherwise social life becomes difficult for us. This type of imitation is the precise opposite of creativity. Imitation is the creation of something that is already in existence. When I was a child, I imitated the way my brother, three years older than myself, ate at the table. My brother sat directly opposite me. He held his chopsticks in his right hand, so I, thinking

I was imitating him exactly, held mine in my left. Once my mother noticed, I changed over to my right hand, but, possibly for this reason, I still wield my chopsticks clumsily even today. I often, in fact, cause amusement by the clumsy way I clutch them in my fist.

I don't mind being laughed at, but what is awkward is when I have a foreign visitor and I take him to some Japanese style restaurant. Since I am often asked to show him how to use chopsticks correctly, I have to take my wife along; she uses them very skillfully, and I tell him to follow her. Before long, I see that he is holding his chopsticks in the proper way whereas I myself am still clumsy. I feel ashamed, but even if I try holding them correctly it doesn't last for more than a minute. Imitation seems to be of no use while one goes on and on repeating it. One never becomes proficient. But just occasionally creativity arises out of this very process of repetition. This kind of imitation becomes, in the broad sense, memory; man has the power of memory, whereby he stores up his experiences.

Without this store of memories, creation is impossible. Yet memory means, as we have seen, to store up experiences and recreate them—repeat them—as required. Recalling is, in itself, a kind of repetition, and there seems to be no element of creation in such a process. Just why it should sometimes give rise to creation is still not really clear.

The Validity of Logic

At some stage or other, human beings acquire the power of rational thought. They learn, for example, to distinguish between truth and falsehood, between right and wrong. They learn that one cannot both affirm and deny the same proposition at the same time, and that one must reject one or the other.

As the power of thought in this sense develops, one becomes able to take in formal logic. We are unconsciously imbued with the idea that one must be faithful to formal logic. Yet in practice we are not faithful. Our reasonings are highly suspect; it often happens that by the time we have developed three or four times the thesis that "such-and-such is so, therefore such-and-such is so," we reach an extremely doubtful conclusion. For in most cases, if one concludes that such-and-such is so with a veracity of, let us say, 70 percent, people will allow themselves to be convinced where that particular fact is concerned. If one then proceeds to deduce from this that such-and-such is so, and if this reasoning, in itself, similarly is 70 percent correct, then the truth of the conclusion will decline to less than 50 percent.

Inductive logic, it is often claimed, is unlike this in being extremely reliable, and it is on inductive logic that scientists rely. Yet this, too, is rather shaky. In many ways inductive logic as it is generally known is very prone to cause misunderstanding. Let us suppose, for example, that one carries out experiments, under all kinds of different conditions, collecting data which lead to the discovery of a law. If one carries out extremely delicate measurements, varying the conditions very frequently and subtly so that one obtains a great deal of data from the experiment—if, in short, the data come to speak for themselves—then nothing very special has been achieved in terms of creativity. In practice, the development of science has been achieved less in this way than by a succession of bold sallies involving the discovery of a simple law from comparatively scanty data and from measurements not in themselves so accurate. To put it round the other way, what is referred to as inductive logic here involves something more than faithfully reproducing experience. There is a leap from the particular to the general. There cannot be discovery until that leap occurs.

At this point the question arises of whether what is referred to as a law in such cases is not really in fact a hypothesis. Both in natural science and in psychology, the fact that something is practically demonstrable is extremely important. However, when one attempts to understand all kinds of empirically observable facts as a whole, it is necessary to resort to laws as intermediaries. The "laws" in question here always have the nature, rather, of hypotheses. When a number of more or less empirical laws are established, physicists tend to find more general laws comprising them, gradually evolving a basic theory. Thus the hypothetical element becomes increasingly pronounced. It is precisely because of this hypothetical quality that demonstration has the decisive importance that it does.

In this connection I stated above that, although we consider formal logic as such to be extremely dependable and are convinced that it is an infallible method of reasoning, in actual use it is not always valid. If, however, one introduces numbers, one can say things with a very high degree of certainty; for the process of demonstrating with numbers, being deductive and formal logic, permits of a perfectly clear-cut yes or no, with nothing intermediate possible. However much one repeats the processes of reasoning, thus, there are no dubious conclusions. Numbers are valid, and it is for this reason everybody puts their faith in them. One way of dealing with numbers, however, is what is known as mathematical induction. To go into details here would create too much of a digression, so the interested reader is referred to Poincaré's *La Science et L'hypothèse.* Very simply speaking, however, it goes as follows: suppose that we have proved that if a statement is true of an arbitrary integer n, then we know that it is true of $n + 1$; and also we have proved that if it is true when n is 1, then it is true of all n's. It is a kind of building up of deductive logic, but first of all the formula in the case of n has to be discovered. Here a creative ability has to be brought into play.

The thing in which mathematics is obviously different from any other branch of study is that in empirical science there must be correspondence with actuality. In modern mathematics, this is not necessary, but in empirical science constant reference to actuality is obligatory. The question arises here of what is actuality, for scholarship would get nowhere if there were nothing but reality as such; we have to understand it with the aid of some intermediary such as words. In practice, we have no way of understanding without an intermediary of some kind. Facts alone are just facts; what is necessary in order to get beyond that point is to put various different facts together to produce something new. In order to express the overall structure of a number of different things—though I dislike using the word "structure," since it is used nowadays in so many different senses—one must introduce concepts and laws.

Creativity lies in this business of putting together to produce something new. This is an extremely important point.

On Analogy

A method in use since ancient times as an aid to understanding something difficult is to draw a parallel with something easier. Generally speaking we never, in practice, understand a thing simply in itself; instead, we draw diagrams, or express it in words, or in a numerical formula. In every case, in short, we are likening it to something else. This, taken in its narrower sense, is the analogy. With an analogy, we understand one side very well, but not the other. The thing we do not understand well seems, somehow, to resemble the one we do. We start to consider just how they are alike—and, in a flash, the thing we did not understand is suddenly elucidated. This kind of thing is happening all the time.

In physics, it takes the form of thinking in terms of models. For example, we take the solar system as a model in considering the structure of the atom. Sometimes, in short, we build up an image by means of which we understand the actual object. We have the ability to form such images, to summon up an organized picture. This ability is a fundamental and vital factor in the manifestation of creativity.

The Lack of Metaphysics

Professor Kikuya Ichikawa of Dōshisha University has distinguished two different types of information which he calls "numerical" and "conceptual" respectively. A typical case of the former is that processed by an electronic computer, which consumes numbers and spews out numbers in return. The information it supplies is "numerical" from first to last; at present, it is incapable of creative activity. It can handle information of a considerable complexity, and carry out difficult calculations for us, but it has no creativity of its own.

One of the reasons for this is that the computer does not possess the ability to conceive, or perceive something as an integral whole. It lacks the human faculty of figurative recognition.

As I pointed out earlier, the present age, seen in terms of the creativity question, is extremely badly off; the display of creativity occurs infrequently when compared to the number of scientists. One reason here, I feel, is that everybody has become "too digital."

Unlike mathematics, physics need to maintain its connections with the actual world in which we live, yet in practice it has become divorced from actual things and preoccupied solely with abstracted theories, experimental data, and the like. The only checking of the facts that is done in practice is to collate experimental data given in numbers and, on the theoretical side, the results of calculating from highly abstract formulae; if the figures on both sides match, people congratulate themselves and that is the end of the matter.

I ask myself just what scientists think they are up to. How can the majority of them be content with such an empty repetition? To put it in different terms, what it comes down to is a lack, a complete lack, of metaphysics. In the nineteenth century, they used to say that science had nothing to do with metaphysics, which must be rigorously excluded; the business of science was to describe and to demonstrate, and it must concern itself with nothing but facts. A similar kind of argument, in a different form, is to be heard today too. But I cannot believe that this is what human learning is all about: it should spring rather from a more basic desire, from what one might call a desire to discover the truth. I have got tired of reiterating this theme, for my opinion seems to be so much at variance with the general trend of our times.

It's something like the man who looks at abstract art and doesn't find it appealing; it is dishonest for him to force himself to nod his head admiringly. Indeed, I very rarely find a work that impresses me; I am sure there are some good works, but I've rarely come across them. It's a great pity; I think there ought to be something more to art than that. However, this is not the time to air my views on art…even so, though, if one reads Schrödinger's *My View of the World* which I mentioned a while ago, one finds the same complaint. It is all very well—he, too, wrote—for physics to be descriptive and demonstrative, but if that meant ridding it of all metaphysics he could not but feel a sense of emptiness. He had this feeling many years ago when physics was much less abstract than it is now.

Identification

I spoke a while ago about analogy. There can be no theory of creativity, I feel, that does not take analogy—which is itself one of the functions of the human intelligence—as its starting point. It is here, and nowhere else, that we must look for the essence of creativity. In actual practice, all the thinkers of the distant past relied on analogy in developing any ideas worthy of the name. Nowadays, I am inclined to think that the fundamental thing in any consideration of creativity is the function of the intelligence known as "identification." I do not mean to imply that identification is in itself the same as creativity; but I do believe that the power of recognition of two things as essentially identical has a decisive importance within the human intellect.

We see a flower, and we know it is a flower; it is not simply that the flower is visible to us, for we can also recognize it for what it is.

As for the nature of recognition: when I say "this is my watch," I am already using the word "watch" as an intermediary, I already know what is the thing known as "watch." I see this watch of mine every day. The memory of it and the actual perception of it by my eyes are identified.

When one reaches a certain age, one begins to wake up to oneself, to become more aware of oneself; one recognizes oneself as oneself; one identifies oneself as oneself. Paradoxically, it is only after one has acquired this awareness of the self that one acquires a philosophy. There is philosophy even before this, of course, but it can be distinguished from what follows as natural philosophy. One is oneself, one is the same as oneself; that appears trivial, and yet, despite that, the identification of oneself as oneself is the beginning of all sophisticated philosophy.

A Manifestation of the Life Force

One of the characteristics that distinguishes the living thing is reproduction. It has the ability to reproduce, to go on recreating, something that is the same, or very similar to, itself, something that even when slightly different is essentially the same. In recent years, it has become clear that an extremely complex molecule known as DNA is the basis of all heredity. This DNA molecule has the ability to produce other molecules the same as itself. Thus if a certain DNA molecule comes into existence it will, given the right environment, create something the same as itself.

The probability of something so extremely particular as a human being coming into existence is basically very small, but once it has come into existence it has the ability to reproduce itself—in other words, it has the mechanism for reproducing itself. Since what occurred by chance can create a large number of things the same as itself, chance ceases to be chance any longer.

Thus one may say that within the basic mechanism of the natural world in which we live, life is also, in itself, a manifestation of a kind of process of identification.

More fundamentally still, there is, at the very bottom of the natural world, something like a kind of invisible mold. This invisible mold is, expressed differently, a law. It produces identical things. It can, for example, produce infinite numbers of electrons each differing in no respect from the rest. That is the essential nature of the mold, and in the reproduction of precisely identical objects we have a process of imitation that is precisely the reverse of creativity in the sense in which we are discussing it here.

If that were all there was to it, there would be no question of creativity, but in the world of living creatures there is what is known as sexual reproduction, in which a new combination is made from the DNAs of male and female. There is also the sudden mutation, where for some reason or other—whether what appears to be chance is really so or not—something different appears suddenly. Where this sudden mutation has the capacity for reproduction, it too goes on increasing indefinitely. At the level of living creatures in general, one may see this process as fortuitous, but in the case of human beings the same kind of thing is done consciously. Human beings consciously produce things that are different. The things that are produced consciously, however, develop, in a different way from living things, an extremely great potential for reproduction.

Such is the case with learning. Tell someone a piece of scientific information, and he perceives the truth of it; in this way, knowledge spreads ever wider and wider. Publish the information in a book, and people read it and are convinced in their turn. An enormous power of reproduction is at work here. Thus conscious effort by human beings at the human level can be seen as an extremely advanced manifestation of the life force, and if this is so then the tenacity that I cited earlier as an extremely important basic factor in the manifestation of creativity can also, I think, be seen as a manifestation of the life force in a very broad sense. Not only things and energy, but all things that actually exist—time, space, our own minds even—are at the same time constantly changing. Changing, becoming—by, for example, the birth of something new—is of their essence.

Fundamentally, this fact is linked up with the way human beings are forever creating new things; thus the display of creativity by a human being is, in the same way, a manifestation of his vitality, of the life force. Tenacity of purpose is also, at bottom, a question of vitality. For such reasons, I am increasingly inclined to see the intellectual ability to identify, which all human beings possess, as a clue to understanding the nature of the world, human beings included; and I am hoping to construct from it some organized theory of creativity. Even if I should succeed, however, I do not persuade myself at it would provide a universal key to the display of creativity.

Things such as creativity just cannot be dealt with in terms of theory alone; without the element of personal experience, in short, it will be flat and lifeless.

However long I go on talking, I seem to reach no final conclusion. If we meet again, I may be able then to present you with a better organized and more advanced theory of creativity, but I hope you have got at least something out of what I have said today.

7

How Scientists Really Think Robert Scott Root-Bernstein

Robert Scott Root-Bernstein is a professor of physiology at Michigan State University and one of the world's authorities on scientific creativity. Having written many articles on the topic and two books, Discovering and Sparks of Genius, his multi-decade research into the careers of hundreds of scientists often distills many of the common characteristics. In this extract from an article he discusses how scientists really think and how this separates many scientists in terms of their success in research.

Here is how the manna is supposed to have arrived according to W.H. Hughes. Hughes was one of Fleming's assistants many years after the event and was therefore most likely an heir to an in-house version of Fleming's research. He says that Fleming had broken his nose very badly as a boy. In consequence, he suffered from an unusual number of colds during the winter. British labs tended to be very cold, too. Thus, and here I quote,

> *One winter day while he was examining some bacterial colonies growing on a culture plate his nose dripped onto some of them. These colonies had not been planted there by Fleming but had come by accident from the air. They were contaminants. The colonies splashed by the mucus melted away; there were others like them on the plate and he was able to sub-culture these onto other plates [so as to perform additional experiments].* [5]

V.D. Allison (I stress the initials, since he must have come in for a lot of ribbing working in a laboratory that specialized in treating patients with venereal diseases) provides us with more details. He was Fleming's assistant at the time Fleming first characterized lysozyme in 1921. He recounts that Fleming,

---

* Reprinted with permission of the publisher from R.S. Root-Bernstein, "How Scientists Really Think," *Perspectives in Biology and Medicine,* Vol. 32 473-488 (1989); extract is pages 476-488.

> *...was busy one evening cleaning up several Petri dishes which had been lying on the bench for perhaps ten days or a fortnight. As he took up one of the dishes in his hand, he looked at it for a long time, showed it to me, and said: "This is interesting." I had a good look at it. It was covered with large yellow colonies which appeared to me to be obvious contaminants. But the remarkable fact was that there was a wide area in which there were no organisms; and another further on, in which the organisms became translucent and glassy. Beyond that, again, were organisms which were in a transitional stage of degradation, between the very glassy ones and those which were fully developed with their normal pigment.*
>
> *Fleming explained that this particular dish was one to which he had added a little of his own nasal mucus, when he had happened to have a cold. This mucus was in the middle of the zone containing no colony. The idea occurred to him that there must be something in the mucus which dissolved or killed the microbes... "Now, that really is interesting," he said again. "We must go into it more thoroughly." His first care was to pick off the organism and stain it by gram. He found it was a large gram-positive coccus, not a pathogen, and not one of the known saprophytic organisms commonly met with, but obviously a contaminating organism which was more likely to have been in the atmosphere of the laboratory and may, of course, have blown through the window.* [6]

Those familiar with the story of Fleming's discovery of penicillin will recognize many similarities.

These two passages from Hughes and Allison give us most of the details of the story, but not all. Gwyn Macfarlane, Fleming's most recent biographer, adds that only a limited number of bacteria are sensitive to lysozyme's activity. Almost all the common pathogens, for example, are unaffected by it. Very few bacteria of any species are as sensitive as the one Fleming dripped on that fateful day in 1921. It is therefore of consequence that the bacterium that contaminated Fleming's petri dish was that particular bacterium and no other [7].

Now, let us analyze this account of Fleming's actions. We have the concurrence of four totally random events. First, Fleming gets a severe cold. Second, at about the same time, he accidentally contaminates a petri dish ostensibly set up for some other purpose. (What that purpose is, we are not told in the chance accounts.) He does not discard the dish despite the contaminating bacterium. Third, the contaminating bacterium just happens to be one that is inordinately sensitive to lysozyme. And finally, Fleming examines the plate and accidentally drips on it, observing the dissolution of the bacterial colonies. What could be a better illustration of the irrationality of the discovery process? It was totally unplanned, unexpected, and so unlikely as to be unrepeatable. There is no induction, since no general rule is drawn from diverse data. There is no deduction, since there is no rule to govern such occurrences. In fact, the existence of the bacteriolysing activity displayed by Fleming's mucus had not been predicted by anyone and probably could not have been. And there is certainly no abduction, since hypothesizing played no role in the discovery. It was a unique event. And yet the result is confirmable.

But wait a minute. Let us look at this from the point of view of a historian of science who is well aware of the philosophical issues. Thomas Kuhn would probably have us ask at this point, what has been discovered, and when? Is the observation of an anomalous activity sufficient to call something a discovery? Must not it also be interpreted, confirmed, and integrated into the existing body of codified scientific knowledge before we call it a discovery [8]? Can one verify or falsify something when one does not even know what the thing is? Looked at this way, this putative discovery seems most unconvincing. Think of it like this: As of that fateful day in November 1921, Fleming could only tell his colleagues that his observations showed some stuff from his nose dissolved an unknown bacterium he had never seen before. Certainly the observation is repeatable, but as any scientist knows, artifacts are repeatable, too—so mere repetition is not sufficient for verification. The result must be interpreted. What does it mean? Lacking any theoretical expectation, does it mean anything? How does he know that his nasal mucus will not dissolve any bacterium whatsoever? How does he know that it is not a chemical effect of pH or of osmotic pressure? How does he know that this particular bacterium, which he has never seen before, is not water soluble? It seems to me the best we can say of the chance account is that it explains not how Fleming discovered lysozyme, but how he discovered the problem that leads to lysozyme.

Even so, the chance account leaves much to be desired-motivation, for example. Why did Fleming repeatedly tell Allison, "This is most interesting?" A thing can be interesting only in relation to some set of expectations. What were Fleming's expectations? An observation can only be testable, in a logical sense, with regard to some explanatory framework. The chance version of the discovery provides neither expectations nor framework.

This is the point at which the irrationalist paradigm fails and the separation between context of discovery and context of justification creates an insuperable barrier to the understanding of scientific research. For in this philosophical context, Fleming's observation has no meaning. He saw something he had never seen before: We all do every day. So what? And yet we know that he devotes the next few months almost solely to developing his observation, and that he reports it as an important discovery. How did he get from the observation to the meaning? What made the observation interesting? To answer that question, we need to get inside Fleming's head.

So, let us start over again and look at Fleming's work through different eyes—eyes that take into account Fleming as a person. For one of the basic flaws in most logic-oriented accounts of discovering is a hidden assumption that anyone in the same position as the discoverer would have seen the same thing and drawn the same conclusions. This is particularly true of chance/irrationalist descriptions of discovery. Thus, the discoverer is often left out of logical accounts of discovery, or is portrayed in such a way that any scientist could replace him. Those who do not perceive the discovery (and could have) are simply said to have suffered a "lost opportunity." To be sure, Allison did not jump up and down shouting, "You've discovered it! Lysozyme! Lysozyme!" Nor could he have. His account clearly exhibits his lack of interest and understanding. This same lack would meet Fleming's initial penicillin work 7 years later, when his colleagues universally commented to one another, "Oh, it's just another one of Flem's little tricks." [7]

Very simply, seeing is not perceiving. We know for a fact that at least five scientists observed the same phenomena Roentgen did, and that they observed them before he did; yet only Roentgen perceived the importance of the scintillations and photographic-plate foggings that led to the discovery of X-rays. We know that at least three of the most eminent crystallographers in Europe had examined the crystals of racemic acid that Louis Pasteur examined in 1848, and that none of them had noticed the asymmetric facets on the crystals that led him to the discovery of molecular asymmetry. Observation is theory directed. We all know this, yet we fail to take the crucial step of realizing that every individual has a different set of theories in their head and a different personality. They will therefore apply what they know and perceive what they see in different ways.

Again, let me define my terms. By personality, I mean the same thing as does the biochemist David Nachmansohn: the sum of the interests, skills, hobbies, experiences, and desires that define the scientist as a human being [9]. All of these will, in Pierre Duhem's words, be rediscovered in the form of the theory the scientist invents [10]. Physicist Gerald Holton has written much the same thing concerning men such as Bohr and Einstein: "there is a mutual mapping of the style of thinking and acting of the genial scientist on the one hand, and the chief unresolved problems of contemporary science on the other." [11] Let us, then, take a look at Fleming as a person in an attempt to understand what unusual styles of thinking he brought to his research, what theories or observations he may have focused on that other bacteriologists did not, and how these may have made him particularly adapted to solving the problem that yields lysozyme.

The most striking thing about Fleming as a person was that he always played games. He was raised in a family that liked to have fun and played everything from poker and bridge to Ping-Pong and quiz games. He would pitch pennies with his landlady when there were no patients to see; play croquet, bowls, bridge, or snooker at his club; shoot competitively and play water polo with the London Scottish Rifle Volunteers (he is even said to have chosen his assistants on their potential for joining the team). He took up golf, but rarely played a straight game-instead he would putt using the club like a snooker cue, or he'd invent whacky rules to make the game "more interesting." He even invented various forms of miniature golf to amuse his friends' children. He was also a practical joker [7]. It would not be amiss to compare him to Richard Feynman.

In short, Fleming was game for anything. He was the only nonartist member of the Chelsea Arts Club and actually took up painting in order to sell a picture so as to qualify for membership. He was not very good, but the point is, he would try almost anything and not take it too seriously. The same philosophy also guided his research. "I play with microbes," he said over and over again in interviews. Sometimes he added, "it's pleasant to break the rules…" [6, 7]

One aspect of this laboratory play manifested itself in the form of germ paintings. Fleming would take microorganisms that developed colors and would "paint" them on a petri plate and incubate it. After a day or two, a "painting" of a ballerina would appear, or a Union Jack, or the fleur-de-lis of the St. Mary's Hospital where he worked. Do not, however, underestimate the skill and knowledge that are required to create such paintings. Not only is there a good amount of hand-eye coordination, there is also a vast store of technical information necessary. Bacteria grow at different rates and develop colors at different temperatures. Some will not grow in the presence of others, or cannot grow at the same pH, or require different nutrients. These germ paintings are therefore "games" requiring the most highly developed bacteriological skills [7].

The paintings also required Fleming constantly to be on the lookout for colored bacteria that he could add to his palette. Because he played with things like germ paintings, he had an unusually wide knowledge of bacterial properties, and he was constantly on the lookout for unusual things. Indeed, his colleagues all said that he had a habit of leaping on unexpected. Allison, for example, reports that when he first arrived in Fleming's lab in 1921, Fleming

> *started to pull my leg about my excessive tidiness. Each evening I put my ' "bench" in order and threw away anything I had no further use for. Fleming told me I was a great deal too careful. He for his part, kept his cultures sometimes for two or three weeks and, before getting*

rid of them, looked very carefully to see whether by chance any unexpected or interesting phenomena had appeared. The sequel was to prove how right he was... [6]

Nor was Fleming alone in his strategy. Max Delbruck taught his students the "principle of limited sloppiness": Be sloppy enough so that something unexpected happens, but not so sloppy you cannot tell what happened. Konrad Lorenz told Desmond Morris in the same vein: "Contrary to your Shakespeare, there is madness in my method!" Morris elaborates that, indeed, "if a group of typical, cautious, orthodox scientists were to make a careful examination of the Lorenzian 'scientific method,' it is unlikely they would ever recover from the experience." [12] Lorenz lived on the verge of chaos with the animals he studied. What these men had learned how to do was to utilize strategies designed to raise problems by purposely breaking the rules of controlled experimentation. I have difficulty reconciling this sort of strategy with what I understand of induction, deduction, or abduction. Yet I do not see that systematic rule breaking as an experimental method for generating problems is any the less a rational operation.

In the same way, Fleming's germ paintings reveal three ways in which he cultivated the unexpected: (1) through play; (2) by developing the breadth of his bacteriological knowledge and the corresponding ability to recognize the unusual; (3) by encouraging unexpected bacterial interactions and contaminations. The paintings also reveal a fourth facet of his style: an ability to arrange phenomena graphically, beautifully, and simply. For it does no good to discover a phenomenon if you cannot display it effectively. This is a skill he learned from his mentor at St. Mary's, Almroth Wright. Wright was a master at inventing simple techniques, such as ways to isolate individual white blood cells with no more than a pipette. Sound impossible, or at best difficult? Not at all. Suck up a drop of blood into the tip of a pipette, allow it to clot, and then blow the clot out. As the blood clots, the white blood cells migrate to the walls of the pipette and stick there when the clot is removed. They can then be washed out of the pipette and examined individually. This simplicity characterized all the techniques Wright pioneered [13].

Fleming quickly picked up his mentor's knack. For example, during World War I it became apparent that, whereas chemical antiseptics worked very well to prevent infection of cuts and abrasions, they were totally ineffective in treating shrapnel wounds, compound fractures, and the like. What was the difference? Fleming decided to find out. The first thing he did was to imagine himself inside the wound. This, again, is a very typical strategy utilized by scientists. Einstein imagined himself riding a light beam [11]; Hannes Alfvén rode each electron and ion trying to imagine the world from its point of view [14]; Peter Debye imagined himself as carbon compounds [15]; Joshua Lederberg [16] and Barbara McClintock [17] imagined they were genes; Jonas Salk imagined how he would act if he were a virus [18]. In effect, they internalized the rules of their science so that public knowledge became indistinguishable from personal knowledge. So Fleming wondered, 'What does a tear wound look like to a bacterium and how does it differ from a cut wound?' Well, tear wounds have irregular surfaces, pocked by nooks and crannies and seams. Cut wounds have nice smooth walls. How could one model this difference?

Fleming invented the spiky test tube [13]—a result of another very unusual skill: he was an excellent glass blower, who not only made his own glassware, but took great joy in making pretty little horses and unicorns for the children of his acquaintances. He therefore had no trouble in making a test tube that had nooks, crannies, and seams in it. To such a test tube, he added bacteria in an appropriate medium. He added the same contents to a regular test tube. He then added chemical antiseptics to both in varying amounts for varying periods of time. The result: a certain minimum amount of antiseptic and time was necessary to kill the bacteria in a regular test tube. No amount of antiseptic or time was adequate to kill off all of the bacteria in a spiky test tube. Somehow the bacteria were able to wall themselves off in the ends of the spikes and survive.

Now, what underlies these examples of graphic display? The operative factors are abstracting and modeling. Abstracting is, by definition, the elimination of everything specific to individual cases to leave only a single, general property. Modeling, on the other hand, denotes the imitation of one thing by another. Models in science are almost always simplifications of a complex situation that cannot be analyzed completely. Most modeling requires abstracting. Fleming's spiky test tube is a valid model of the interior of a tear wound only insofar as it correctly represents one specific element of such a wound. Again, as with problem-generation-through-rule-breaking, I cannot reconcile abstracting and modeling with induction, deduction, or abduction. One does not induce, deduce, or abduce abstractions or models. One utilizes other methods, such as the successive elimination of properties until only the ones required to display the phenomenon of interest are left. In other words, one does not reason from principles or from data-one discards as many of both as possible.

So, we have Fleming the game player, actively looking for unusual bacteria, especially for brightly colored ones like the unknown yellow coccus he finds on his petri dish in 1922. He has sufficiently diverse knowledge to know what properties to expect from such a bacterium, and he has an unusual ability to make such bacteria display these properties. But had he dripped nasal mucus onto bacteria so many times before that he actually knew that the dissolution of the bacterium was unusual? Did he, in fact, drip accidentally?

Here we reach another problem with the chance accounts. Hughes reports that Fleming had attached a leather drip guard to his microscope specifically to prevent just such accidents from occurring [5]. Moreover, Fleming states in the published version of his paper on lysozyme that

> *The lysozyme was first noticed during some investigations made on a patient suffering from acute coryza [a cold]. The nasal secretion of this patient was cultivated daily on blood agar plates, and for the first three days of the infection there was no growth, with the exception of an occasional staphylococcus colony. The culture made from the nasal mucus on the fourth day showed in 24 hours a large number of small colonies which, on examination, proved to be large grampositive cocci...* [19]

In short, Fleming claims that the experiment was done on purpose that the yellow bacterium Allison thought was a contaminant was actually isolated from someone's nose. So who is this patient, and why cultivate his or her nasal secretions?

Fleming's lab notebooks exist for this period. There is no doubt who the patient is: it is Fleming himself. On November 21, 1921, Fleming draws in his lab notebook a picture of a petri plate with a colony marked "Staphyloid coccus from A.F.'s nose." [7] A.F. is, of course, Alexander Fleming. So, Fleming did have a cold as Allison and Hughes reported. That is important. He did not, however, drip accidentally; he dripped on purpose. The question, then, is for what purpose?

Before I address the question of purpose, let me first stress several points. Fleming did not isolate his lysozyme-sensitive bacterium by accident. He was looking for something like it. How do we know? Because both his written report and his notebooks say that he isolated a number of other bacteria on previous days of the cold, but these were common pathogens and did not interest him. He continued to isolate bacteria until he found one that did interest him. In short, he had criteria for recognizing what he was looking for. Since the criteria of choice predate his experiment, did he deduce them from some prior rule? What were his criteria of choice? Where did he derive them from? And why was he doing the experiment in the first place?

All three questions have the same answer, and again, that answer is to be found in Fleming's notebooks. Oddly, although the facts of the case are known, previous historians and scientists, such as Macfarlane who has studied the notebooks in detail, have not been able to make sense of them. Their failure, in my view, has been to stick to the "facts" when the problem can only be solved by introducing Fleming the fun-loving game player.

What the notebooks show is that Fleming was looking for bacteriophage—that is, a virus that "eats" bacteria. How do we know? Because beginning several days before and extending for over a month afterwards, every notebook entry relating to what we now call lysozyme was headed with the title: "Bacteriophage." Not only did Fleming initiate his experiments to look for bacteriophage, he actually thought for a time he had found it. How can that be? And how does it explain the genesis of lysozyme?

To understand the connection, one must know a few things about the state of virus studies in 1921. Filterable, infectious particles that destroyed bacteria were first discovered in 1915 by Felix d'Hérelle in the diarrhea of locusts (one of the odder facts of the history of science!) and by Frederick Twort at about the same time. A year or so later, during World War I, d'Hérelle isolated another bacteria-eating, filterable agent that "ate" gut bacteria from the stools of patients suffering from dysentery. The important point is that all forms of bacteriophage identified by 1921 were associated with what were colloquially known as "the runs."

Enter Fleming the mischievous game player. His problem: What causes his frequent and uncomfortable runny noses? Wait a minute! Runny bottoms are caused by bacteriophage infections! Why not runny noses!? A hypothesis is born of verbal analogy!

Ah! You object. I am being too silly. I cannot be serious. But I am. Arthur Koestler suggested long ago that the structure of discovery was similar to the structure of jokes [20]. A number of philosophers of science-for example, Peter Caws [21] and Errol Harris [22] have suggested that scientific reasoning will turn out to be merely a form of everyday reasoning. And why not? Are not scientists human beings? Did not Diderot comment in his Dream of d'Alembert that, "ideas awaken each other and they do so because they have always been related"? Have not Mitchell Wilson [23] and Thomas Hughes [24] written that the scientist and inventor is characterized by an intense, almost poetical awareness of words and their meanings; of metaphor and analogy? Have you ever met a scientist who did not pun? So why not ideas awakening one another because they are couched in the same terms? Why should not a play on words, a pun, in the hands of a master of play such as Fleming, be just as effective in generating scientific hypotheses as induction or deduction, with the added benefit of making the process surprising and fun?

Besides, Fleming is not alone in generating hypotheses through verbal punning or recognition of similar terminology. Svante Arrhenius invented his concept of ionic dissociation —that salts dissolve into electrically charged ions in solution-through such a similarity in language. Studies of mass-action effects and electrolysis in solutions had led scientists to talk of "active" and "inactive" molecules. Independently, investigators studying the effect of light on rates of reactions in gases—the so-called photochemical effect—also began speaking of "active" and "inactive" molecules. This similarity of language led Arrhenius to link the two disparate fields [25]. Also striking is the example of botanist Ralph Lewis who was studying the dissemination of fungal spores by flies. Repeated experimental failures led him to analyze the differences between his setup and the natural situation. He quickly realized that, whereas he used dried spores, spores in nature are found in a substance called "honeydew." As soon as he saw the word "honeydew," it flashed through his mind, "Would honey do?" It did [26]. Thomas Hughes has documented several similar cases in the field of technological invention [24] and who knows how many similar stories have remained in the domain of private recollection.

In the event, the runny-bottom-runny-nose hypothesis is the only one that can explain Fleming's behavior. It sets his expectations and dictates his actions. By looking for a bacteriophage, Fleming preset his mind to look for particular kinds of activity. One should be able to mimic d'Hérelle's and Twort's discoveries of bacteriophage by first isolating an appropriate bacterium and then demonstrating the existence of a filterable agent capable of "eating" it. The classic bacteriophage experiment is to grow one's bacterium in solution until the solution is cloudy. One then adds the material thought to contain the bacteriophage and within a few minutes, the solution will quite suddenly become clear as the viruses burst out of the bacteria, causing them to lyse. It is a very graphic demonstration of the sort Fleming must have appreciated. The analogy to what he had done on the petri dish is clear. And to make it clearer, the very next experiment Fleming did was to grow his bacterium in solution and then add nasal mucus. As expected, the solution grew suddenly clear. The next month was spent characterizing the new "bacteriophage" by filtering it, heating it, determining its susceptibility to chemical degradation, and so forth [7].

That is when things started to go "wrong." One test for bacteriophage activity is to take a small sample of the now-clarified bacterial solution and to demonstrate that it has at least as much, if not more, bacteriophage activity as the original sample. Since bacteriophage replicate, one can dilute the material quite far and still have it return to its original potency after a few hours. Not so with an enzyme. Enzymes do not replicate. The more you dilute them, the less activity you observe. Moreover, bacteriophages and enzymes react somewhat differently to heat and pH. Fleming soon realized he had found an enzyme, not a virus, and he never reported why he undertook the experiments. Subsequent chance accounts are therefore based on an argument from ignorance.

There are two points to be stressed here. The first is that in making his analogy between colds and diarrhea, Fleming imports wholesale the entire corpus of techniques, tests, and criteria already developed for bacteriophage studies into a new realm. He thinks he is extending an existing corpus of science to embrace a wider field of phenomena. Viruses may, he thinks, cause not only intestinal disease, but the common cold. (Notice he was not too far wrong!) Therefore, the logic of discovery, or more accurately the logic of research, is identical and inseparable from the logic of verification. The test of one idea becomes the discovery of something else and the initial hypothesis is neither verified nor falsified. It becomes irrelevant. Surprise! This is my solution to the paradox raised by separating the logic of discovery from the logic of justification. In my view, scientific discovery cannot be so divided. Rather, there is a logic of research that leads to the surprise of discovery through the recognition of unexpected problems generated by the research itself.

The second point I want to make-albeit in passing-is that virtually every so-called chance discovery that has been reanalyzed in light of newly found manuscript evidence has yielded the same sort of story I have just given here. Minkowski's discovery of diabetes, several of Pasteur's classic "chance" discoveries, Fleming's work on penicillin, Richet's discovery of anaphylaxis, and so forth, all turn out to be due to testing incorrect hypotheses generated by extension of existing observations. In each case a preexisting expectation was not fulfilled, and the anomalous result required rethinking of that expectation [27]. In consequence, the genesis of the idea was never reported and scientists, historians, and philosophers have been left with the incorrect impression that the research occurred "by chance." In each case, records of the actual research belie these irrationalist accounts. I therefore find no basis in history of science-or for that matter, in my own research-for separating the context of discovery from the context of justification. They are one and the same. I find no evidence that scientists act irrationally or illogically in setting up research programs. On the contrary, the best specifically utilize strategies designed to maximize their probability of encountering problems.

Several conclusions follow from this analysis. First, I want to propose a complete shift in the way we view science. Philosophers (and most scientists) tend to view science as the "Search for Solutions," as Horace Judson calls it [16]. Thus, philosophers tend to focus solely on the propositional aspects of science. In contrast, working scientists spend the majority of their time raising problems—not solving them. That was the whole point of Fleming's insistence on leaving things lying about to provoke the unexpected, Lorenz's "madness," Delbruck's "controlled sloppiness." These are not ways of solving problems; they are ways of raising them. Hence, we must alter our view of science accordingly and view it as the "Quest for Questions." Indeed, it seems to me that unless we alter our perspective in this way, we cannot understand the multitude of statements by men such as Bohr, Heisenberg, J.J. and G.P. Thomson, Santiago Ramón y Cajal, Charles Richet, and so many others that, and here I quote Einstein and Infeld:

> *The formulation of a problem is often more essential than its solution, which may be merely a matter of mathematical or experimental skill. To raise new questions, new possibilities, to regard old problems from a new angle, requires creative imagination and works real advance in science.* [28]

It is not science's ability to reach solutions that needs to be accounted for, but its ability to define solvable problems.

Second, induction, deduction, and abduction will not suffice to solve the range of problems scientists address. Scientific "tools of thought" are more diverse than this and have been developed not only to reason and to test, but to invent. These tools of invention include (and are probably not limited to): abstracting, modeling, analogizing, pattern forming and pattern recognition, aesthetics, visual thinking, playacting, and manipulative skill. I suggest that scientists might better educate their successors if they included these tools of thought in teaching their science [27, 29, 30].

Third, along with these extralogical tools of thought, we must also recognize that scientists employ strategies of research aimed at increasing their probability of discovering. Fleming, for example, invented his bacterial paintings as a strategy for finding odd bacteria. He also, like many other scientists, had a tendency to break rules just to see what happens. Such strategies—others, such as "turning it on its head," or "push it till it fails," quickly come to mind-might be codified, just as they are in chess or war, and thus developed into important additions to our understanding of science [27].

Finally, we must accept that science is done by individuals, each of whom possesses a different style of research and each of whom is heir to different traditions of research. Every scientist is unique in his personality and, hence, unique in his perceptions. Thus, scientists seeing the same thing may not perceive the same thing. I stress that this conclusion is a logical correlate of the already widely accepted proposition that observation is theory dependent. Recognition of this individual variability does not preclude the possibility of a logical analysis of science. It simply means that such an analysis will have to be done on the pattern of evolutionary analyses of nature-that is to say, with an eye to variability as an essential element in the process. In summary, then, I do not believe that the traditional forms of logical analysis can explain how scientists reason and behave. We need a new kind of analysis that can account for problem invention, playfulness, fun, punning, unexpected surprises, analogizing, modeling, and the like. We need, in short, to take a less serious and more subjective view of the scientific process that puts traditional logic in its (limited) place:

The seduction of induction
Is that data yield what's true;
And production by deduction
Can be done when facts are few.
Yet presumptions and assumptions
False invalidate one's reason:
In reality, factuality
May be logic's hidden treason.
I advise instead analogizin';
Pattern forming's neat.

The attraction of abstraction
Is its simplifying feat.
He who coddles newborn models
Helps raise theories as he ought;
Play and punning, just-for-funning,
Can yield most surprising thought.

REFERENCES

[1] De la Torre, L., *Dr. Sam Johnson, Detector.* (New York: Knopf, 1946).
[2] Popper, K. *The Logic of Scientific Discovery.* (New York: Basic, 1959).
[3] Pierce, C.S., *Collected Papers.* (Cambridge, Mass.: Harvard Univ. Press, 1931).
[4] Poincaré, H. "Mathematical creation". In: *The Foundation of Science*, Translated by B.B. Halstead (New. York: Science, 1913). Ed. note-see chapter 13 of this volume.
[5] Hughes, W.H., *Alexander Fleming and Penicillin.* (London: Priory, 1974).
[6] Maurois, A., *The Life of Sir Alexander Fleming*, Translated by G. Hopkins (New York: Dutton, 1959).
[7] MacFarlane, G. *Alexander Fleming: The Man and the Myth.* (Cambridge, Mass.: Harvard Univ. Press, 1984).
[8] Kuhn, T.S., *The Essential Tension.* (Chicago: Univ. Chicago Press, 1977).
[9] Nachmansohn, D. "Biochemistry as part of my life." *Annual Review of Biochemistry.* Vol. 4, 1 (1972).
[10]Lowinger, A., *The Methodology of Pierre Duhem.* (New York: Columbia Univ. Press, 1941).
[11] Holton, G., *Thematic Origins of Scientific Thought: Kepler to Einstein.* (Cambridge, Mass.: Harvard Univ. Press, 1973).
[12] Morris, D. *Animal Days.* (New York: Bantam, 1979).
[13] Colebrook, L., *Almroth Wright: Provocative Doctor and Thinker.* (London: Heinemann, 1954).
[14] Alfvén, H., "Memoirs of a dissident scientist." *American Scientist,* Vol. 76 249 (1988). Ed. note see chapter 18 of this book.

[15] Debye, P. "Peter J.W. Debye [Interview]." In: *The Way of the Scientist*, edited by the editors of International Science and Technology. (New York: Simon & Schuster, 1966).
[16] Judson, H.F., *The Search for Solutions.*(New York: Holt, Rinehart & Winston, 1980).
[17] Keller, E.F. *A Feeling for the Organism: The Life and Work of Barbara McClintock.* (San Francisco: W.H. Freeman, 1983).
[18] Salk, J. *The Anatomy of Reality.* (New York: Columbia Univ. Press, 1983).
[19] Fleming, A. "On a remarkable bacteriolytic substance found in secretions and tissues". *Proceedings of the Royal Society of London* Vol. 93B:306 (1922).
[20] Koestler, A., *The Act of Creation.* (London: Hutchinson, 1976).
[21] Caws, P., "The structure of discovery." *Science* Vol. 166, 1374 (1969).
[22] Harris, E.E., *Hypothesis and Perception: The Roots of Scientific Method.* (London: Allen & Unwin, 1970).
[23] Wilson, M. *Passion to Know.* (Garden City, N.Y.: Doubleday, 1972).
[24] Hughes, T.P. "How did the heroic inventors do it?" *American Heritage of Invention & Technology*, 18, Fall 1985.
[25] Root-Bernstein, R.S. *The ionists: founding physical chemistry, 1872-1890.* Ph.D. diss., Princeton University, 1980; Ann Arbor, Mich.: Univ. Microfilms, 1981.
[26] Lewis, R., *The field inoculation of rye with Cliviceps purpurea (Fr. Tul.).* Ph.D. diss., Michigan State College of Agriculture (now Michigan State University), 1944.
[27] Root-Bernstein, R.S., *Discovering.* (Cambridge, Mass.: Harvard Univ. Press, 1989).
[28] Einstein, A. & Infeld, L., *The Evolution of Physics.* (New York: Simon & Schuster, 1938).
[29]Root-Bernstein, R.S., "Visual thinking: the art of imagining reality." *Transactions of the American Philosophical Society,* Vol. 75(Pt. 6) 50 (1985).
[30] Root-Bernstein, R.S., "Harmony and beauty in medical research." *Journal of Molecular and Cellular Cardiology,* Vol. 19, 1043 (1987).

8

The Art of Scientific Investigation: Chapter 5: Imagination William Ian Beardmore Beveridge

"With accurate experiment and observation to work upon, imagination becomes the architect of physical theory."
—Tyndall

Productive thinking

This chapter and the next contain a brief discussion on how ideas originate in the mind and what conditions are favourable for creative mental effort. The critical examination of the processes involved will be rendered easier if I do as I have done in other parts of this book, and make an arbitrary division of what is really a single subject. Consequently much of the material in this chapter should be considered in connection with Intuition and much of the next chapter applies equally to Imagination.

Dewey analyses conscious thinking into the following phases. First comes awareness of some difficulty or problem which provides the stimulus. This is followed by a suggested solution springing into the conscious mind. Only then does reason come into play to examine and reject or accept the idea. If the idea is rejected, our mind reverts to the previous stage and the process is repeated. The important thing to realise is that the conjuring up of the idea is not a deliberate, voluntary act. It is something that happens to us rather than something we do [1].

In ordinary thinking ideas continually "occur" to us in this fashion to bridge over the steps in reasoning and we are so accus-

* Reprinted from W.I.B. Beveridge, The Art of Scientific Investigation (New York: W.W. Norton, 1957) 53-67 under Creative Commons Attribution-ShareAlike License 3.0

tomed to the process that we are hardly aware of it. Usually the new ideas and combinations result from the immediately preceding thought calling up associations that have been developed in the mind by past experience and education. Occasionally, however, there flashes into the mind some strikingly original idea, not based on past associations or at any rate not on associations that are at first apparent. We may suddenly perceive for the first time the connection between several things or ideas, or may take a great leap forward instead of the usual short step where the connections between each pair or set of ideas are well established and "obvious." These sudden, large progressions occur not only when one is consciously puzzling the problem but also not uncommonly when one is not thinking of anything in particular, or even when one is mildly occupied with something different, and in these circumstances they are often startling. Although there is probably no fundamental difference between these ideas and the less exciting ones that come to us almost continually, and it is not possible to draw any sharp distinction, it will be convenient to consider them separately in the next chapter under the title "intuitions." In this section we will draw attention to some general features of productive or creative thinking.

Dewey advocates what he calls "reflective thinking," that is, turning a subject over in the mind and giving it ordered and consecutive consideration, as distinct from the free coursing of ideas through the head. Perhaps the best term for the latter is day-dreaming; it also has its uses, as we shall see presently. But thinking may be reflective and yet be inefficient. The thinker may not be sufficiently critical of ideas as they arise and may be too ready to jump to a conclusion, either through impatience or laziness. Dewey says many people will not tolerate a state of doubt, either because they will not endure the mental discomfort of it or because they regard it as evidence of inferiority.

> *To be genuinely thoughtful, we must be willing to sustain and protract that state of doubt which is the stimulus to thorough inquiry, so as not to accept an idea or make a positive assertion of a belief, until justifying reasons have been found.* [1]

Probably the main characteristic of the trained thinker is that he does not jump to conclusions on insufficient evidence as the untrained man is inclined to do.

It is not possible deliberately to create ideas or to control their creation. When a difficulty stimulates the mind, suggested solutions just automatically spring into the consciousness. The variety and quality of the suggestions are functions of how well prepared our mind is by past experience and education pertinent to the particular problem. What we can do deliberately is to prepare our minds in this way, voluntarily direct our thoughts to a certain problem, hold attention on that problem and appraise the various suggestions thrown up by the subconscious mind. The intellectual element in thinking is, Dewey says, what we do with the suggestions after they arise.

Other things being equal, the greater our store of knowledge, the more likely it is that significant combinations will be thrown up. Furthermore, original combinations are more likely to come into being if there is available a breadth of knowledge extending into related or even distant branches of knowledge. As Dr. E.L. Taylor says:

> *New associations and fresh ideas are more likely to come out of a varied store of memories and experience than out of a collection that is all of one kind.* [2]

Scientists who have made important original contributions have often had wide interests or have taken up the study of a subject different from the one in which they were originally trained. Originality often consists in finding connections or analogies between two or more objects or ideas not previously shown to have any bearing on each other.

In seeking original ideas, it is sometimes useful to abandon the directed, controlled thinking advocated by Dewey and allow one's imagination to wander freely—to day-dream. Harding says all creative thinkers are dreamers. She defines dreaming in these words:

> *Dreaming over a subject is simply...allowing the will to focus the mind passively on the subject so that it follows the trains of thought as they arise, stopping them only when unprofitable but in general al-*

lowing them to form and branch naturally until some useful and interesting results occur. [3]

Max Planck said:

> *Again and again the imaginary plan on which one attempts to build up order breaks down and then we must try another. This imaginative vision and faith in the ultimate success are indispensable. The pure rationalist has no place here.* [4]

In meditating thus, many people find that visualising the thoughts, forming mental images, stimulates the imagination. It is said that Clerk Maxwell developed the habit of making a mental picture of every problem. Paul Ehrlich was another great advocate of making pictorial representations of ideas, as one can see from his illustrations of his side-chain theory. Pictorial analogy can play an important part in scientific thinking. This is how the German chemist Kekulé hit on the conception of the benzene ring, an idea

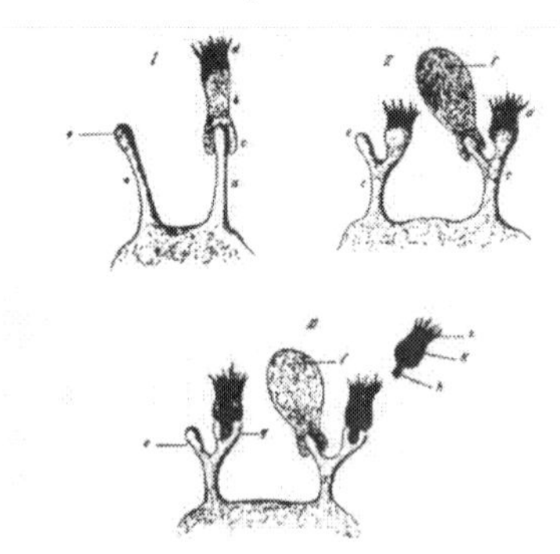

Ehrlich's drawings of his side-chain theory

that revolutionised organic chemistry. He related how he was sitting writing his chemical text-book:

> *But it did not go well; my spirit was with other things. I turned the chair to the fireplace and sank into a half sleep. The atoms flitted before my eyes. Long rows, variously, more closely, united; all in movement wriggling and turning like snakes. And see, what was that? One of the snakes seized its own tail and the image whirled scornfully before my eyes. As though from a flash of lightning I*

> *awoke; I occupied the rest of the night in working out the consequences of the hypothesis...Let us learn to dream, gentlemen.* [5]

However, physics has reached a stage where it is no longer possible to visualise mechanical analogies representing certain phenomena which can only be expressed in mathematical terms.

In the study of infectious diseases, it is sometimes helpful to take the biological view, as Burnet has done, and look upon the causal organism as a species struggling for continued survival, or even, as Zinsser has felt inclined to do with typhus, which he spent a lifetime studying, personifying the disease in the imagination.

An important inducement to seeking generalisations, especially in physics and mathematics, is the love of order and logical connection between facts. Einstein said:

> *There is no logical way to the discovery of these elemental laws. There is only the way of intuition, which is helped by a feeling for the order lying behind the appearance.* [6]

W.H. George remarks that a feeling of tension is produced when an observer sees the objects lying in his field of vision as forming a pattern with a gap in it, and a feeling of relaxation or satisfaction is experienced when the gap is closed, and all parts of the pattern fit into their expected places. Generalisations may be regarded as patterns in ideas [7]. Another phenomenon which may be explained by this concept is the satisfaction experienced on the completion of any task. This may be quite unassociated with any consideration of reward for it applies equally to unimportant, self-appointed tasks such as doing a crossword puzzle, climbing a hill or reading a book. The instinctive sense of irritation we feel when someone disagrees with us or when some fact arises which is contrary to our beliefs may be due to the break in the pattern we have formed.

The tendency of the human mind to seek order in things did not escape the penetrating intelligence of Francis Bacon. He warned against the danger that this trait may mislead us into believing we see a greater degree of order and equality than there really is.

When one has succeeded in hitting upon a new idea, it has to be judged. Reason based on knowledge is usually sufficient in everyday affairs and in straightforward matters in science, but in research there is often insufficient information available for effective reasoning. Here one has to fall back on "feelings" or "taste." Harding says:

> *If the scientist has during the whole of his life observed carefully, trained himself to be on the look out for analogy, and possessed himself of relevant knowledge, then the 'instrument of feeling'...will become a powerful divining rod...in creative science feeling plays a leading part.* [3]

Writing of the importance of imagination in science Tyndall said:

> *Newton's passage from a falling apple to a falling moon was an act of the prepared imagination. Out of the facts of chemistry the constructive imagination of Dalton formed the atomic theory. Davy was richly endowed with the imaginative faculty, while with Faraday its exercise was incessant, preceding, accompanying and guiding all his experiments. His strength and fertility as a discoverer are to be referred in great part to the stimulus of the imagination.* [8]

Imagination is of great importance not only in leading us to new facts, but also in stimulating us to new efforts, for it enable us to see visions of their possible consequences. Facts and ideas are dead in themselves and it is the imagination that gives life to them. But dreams and speculations are idle fantasies unless reason turns them to useful purpose. Vague ideas captured on flights of fancy have to be reduced to specific propositions and hypotheses.

False trails

While imagination is the source of inspiration in seeking new knowledge, it can also be dangerous if not subjected to discipline; a fertile imagination needs to be balanced by criticism and judgment. This is, of course, quite different from saying it should be repressed or crushed. The imagination merely enables us to wander into the darkness of the unknown where, by the dim light of the knowledge that we carry, we may glimpse something that seems of interest. But when we bring it out and examine it more closely it usually proves to be only trash whose glitter had caught our attention. Things not clearly seen often take on grotesque forms. Imagination is at once the source of all hope and inspiration but also of frustration. To forget this is to court despair.

Most hypotheses prove to be wrong whatever their origin may be. Faraday wrote:

> *The world little knows how many of the thoughts and theories which have 'passed through the mind of a scientific investigator have been crushed in silence and secrecy by his own severe criticism and adverse examinations; that in the most successful instances not a tenth of the suggestions, the hopes, the wishes, the preliminary conclusions have been realised.*

Every experienced research worker will confirm this statement.

Darwin went even further:

> *I have steadily endeavoured to keep my mind free so as to give up any hypothesis, however much beloved (and I cannot resist forming one on every subject) as soon as facts are shown to be opposed to it…cannot remember a single first formed hypothesis which had not after a time to be given up or be greatly modified.* [9]

T.H. Huxley said that the great tragedies of science are the slaying of beautiful hypotheses by ugly facts. F.M. Burnet has told me that most of the "bright ideas" that he gets prove to be wrong.

There is nothing reprehensible about making a mistake, provided it is detected in time and corrected. The scientist who is excessively cautious is not likely to make either errors or discoveries. Whitehead has expressed this aptly: "panic of error is the death of progress." Humphrey Davy said:

> *The most important of my discoveries have been suggested to me by my failures.*

The trained thinker shows to great advantage over the untrained person in his reaction to finding his idea to be wrong. The former profits from his mistakes as much as from his successes. Dewey says:

> *What merely annoys and discourages a person not accustomed to thinking…is a stimulus and guide to the trained enquirer.…It either brings to light a new problem or helps to define and clarify the problem.* [1]

The productive research worker is usually one who is not afraid to venture and risk going astray, but who makes a rigorous test for error before reporting his findings. This is so not only in the biological sciences but also in mathematics. Hadamard states that good mathematicians often make errors but soon perceive and correct them, and that he himself makes more errors than his students. Commenting on this statement, Sir Frederic Bartlett, Professor of Psychology at Cambridge, suggests that the best single measure of mental skill may lie in the speed with which errors are detected and thrown out [10]. Lister once remarked:

> *Next to the promulgation of the truth, the best thing I can conceive that a man can do is the public recantation of an error.*

W.H. George points out that even with men of genius, with whom the birth rate of hypotheses is very high, it only just manages to exceed the death rate.

Max Planck, whose quantum theory is considered by many to be an even more important contribution to science than Einstein's theory of relativity, said when he was awarded the Nobel Prize:

> *Looking back...over the long and labyrinthine path which finally led to the discovery [of the quantum theory], I am vividly reminded of Goethe's saying that men will always be making mistakes as long as they are striving after something.* [4]

Einstein in speaking of the origin of his general theory of relativity said:

> *These were errors in thinking which caused me two years of hard work before at last, in 1915, I recognised them as such....The final results appear almost simple; any intelligent undergraduate can understand them without much trouble. But the years of searching in the dark for a truth that one feels, but cannot express; the intense desire and the alternations of confidence and misgiving, until one breaks through to clarity and understanding, are only known to him who has himself experienced them.* [6]

Perhaps the most interesting and revealing anecdote on these matters was written by Hermann von Helmholtz [11]:

> *In 1891 I have been able to solve a few problems in mathematics and physics including some that the great mathematicians had puzzled over in vain from Euler onwards...But any pride I might have felt in my conclusions was perceptibly lessened by the fact that I knew that the solution of these problems had almost always come to me as the gradual generalisation of favourable examples, by a series of for-*

tunate conjectures, after many errors. I am fain to compare myself with a wanderer on the mountains who, not knowing the path, climbs slowly and painfully upwards and often has to retrace his steps because he can go no further—then, whether by taking thought or from luck, discovers a new track that leads him on a little till at length when he reaches the summit he finds to his shame that there is a royal road, by which he might have ascended, had he only had the wits to find the right approach to it. In my works, I naturally said nothing about my mistake to the reader, but only described the made track by which he may now reach the same heights without difficulty.

Curiosity as an incentive to thinking

In common with other animals we are born with an instinct of curiosity. It provides the incentive for the young to discover the world in which they live—what is hard or soft, movable or fixed, that things fall downwards, that water has the property we call wetness, and all other knowledge required to enable us to accommodate ourselves to our environment. Infants whose mental reflexes have not yet been conditioned are said not to exhibit the "attack-escape" reaction as do adults, but to show rather the opposite type of behaviour. By school age we have usually passed this stage of development, and most of our acquisition of new knowledge is then made by learning from others, either by observing them or being told or reading. We have gained a working knowledge of our environment and our curiosity tends to become blunted unless it is successfully transferred to intellectual interests.

The curiosity of the scientist is usually directed toward seeking an understanding of things or relationships which he notices have no satisfactory explanation. Explanations usually consist in connecting new observations or ideas to accepted facts or ideas. An explanation may be a generalisation which ties together a bundle of data into an orderly whole that can be connected up with current knowledge and beliefs. That strong desire scientists usually have to seek underlying principles in masses of data not obviously related may be regarded as an adult form or sublimation of curiosity. The student attracted to research is usually one who retains more curiosity than usual.

We have seen that the stimulus to the production of ideas is the awareness of a difficulty or problem, which may be the realisation of the present unsatisfactory state of knowledge.

People with no curiosity seldom get this stimulus, for one usually becomes aware of the problem by asking why or how some process works, or something takes the form that it does. That a question is a stimulus is demonstrated by the fact that when someone asks a question it requires an effort to restrain oneself from responding.

Some purists contend that scientists should wonder "how" and not "why." They consider that to ask "why" implies that there is an intelligent purpose behind the design of things and that activities are directed by a supernatural agency toward certain aims. This is the teleological view and is rejected by present-day science, which strives to understand the mechanism of all natural phenomena. Von Bruecke once remarked:

> *Teleology is a lady without whom no biologist can live; yet he is ashamed to show himself in public with her.*

In biology, asking "why" is justified because all events have causes; and because structures and reactions usually fulfill some function which has survival value for the organism, and in that sense they have a purpose. Asking "why" is a useful stimulus towards imagining what the cause or purpose may be. "How" is also a useful question in provoking thought about the mechanism of a process.

There is no satisfying the scientists' curiosity, for with each advance, as Pavlov said, "we reach a higher level from which a wider field of vision is open to us, and from which we see events previously out of range." It may be appropriate to give here an illustration of how curiosity led John Hunter to carry out an experiment which led to an important finding.

While in Richmond Park one day Hunter saw a deer with growing antlers. He wondered what would happen if the blood supply were shut off on one side of the head. He carried out the experiment of tying the external carotid artery on one side, whereupon the corresponding antler lost its warmth and ceased to grow. But after a while the horn became warm again and grew. Hunter ascertained that his ligature still held, but neighbouring arteries had increased in size till they carried an adequate supply of blood. The existence of collateral circulation and the possibility of its increasing were thus discovered. Hitherto no one had dared to treat aneurism by ligation for fear of gangrene, but now Hunter saw the possibilities and tried ligation in the case of popliteal aneurism. So the Hunterian operation, as it is known in surgery today, came into an assured existence. An insatiable curiosity seems to have been the driving force behind Hunter's prolific mind which laid the foundation of modern surgery. He even paid the expenses of a surgeon to go and observe whales for him in the Greenland fisheries.

Discussion as a stimulus to the mind

Productive mental effort is often helped by intellectual intercourse. Discussing a problem with colleagues or with lay persons may be helpful in one of several ways.

(a) The other person may be able to contribute a useful suggestion. It is not often that he can help by directly indicating a solution of the impasse, because he is unlikely to have as much pertinent knowledge as has the scientist working on the problem, but with a different background of knowledge he may see the problem from a different aspect and suggest a new approach. Even a layman is sometimes able to make useful suggestions. For example, the introduction of agar for making solid media for bacteriology was due to a suggestion of the wife of Koch's colleague Hesse [13].

(b) A new idea may arise from the pooling of information or ideas of two or more persons. Neither of the scientists alone may have the information necessary to draw the inference which can be obtained by a combination of their knowledge.

(c) Discussion provides a valuable means of uncovering errors. Ideas based on false information or questionable reasoning may be corrected by discussion and likewise unjustified enthusiasms may be checked and brought to a timely end. The isolated worker who is unable to talk over his work with colleagues will more often waste his time in following a false trail.

(d) Discussion and exchange of views is usually refreshing, stimulating and encouraging, especially when one is in difficulties and worried.

(e) The most valuable function of discussion is, I believe, to help one to escape from an established habit of thought which has proved fruitless, that is to say, from conditioned thinking. The phenomenon of conditioned thinking is discussed in the next section.

Discussions need to be conducted in a spirit of helpfulness and mutual confidence and one should make a deliberate effort to keep an open receptive mind. Discussions are usually best when not more than about six are present. In such a group no one should be afraid of admitting his ignorance on certain matters and so having it corrected, for in these days of extreme specialisation everyone's knowledge is restricted. Conscious ignorance and intellectual honesty are important attributes for the research man. Free discussion requires an atmosphere unembarrassed by any suggestion of authority or even respect. Brailsford Robertson tells the story of the great biochemist, Jacques Loeb, who, when asked a question by a student after a lecture, replied characteristically:

> *I cannot answer your question, because I have not yet read that chapter in the text-book myself, but if you will come to me tomorrow I shall then have read it, and may be able to answer you.* [14]

Students sometimes quite wrongly think that their teachers are almost omniscient, not knowing that the lecturers usually spend a considerable amount of time preparing their lectures, and that outside the topic of the lecture their knowledge is often much less impressive. Not only does the author of a text-book not carry in his head all the information in the book, but the author of a research paper not infrequently has to refer to the paper to recall the details of the work which he himself did.

The custom of having lunch and afternoon tea in groups at the laboratory is a valuable one as it provides ample opportunities for these informal discussions. In addition, slightly more formal seminars or afternoon tea meetings at which workers present their problems before and during, as well as after, the investigation are useful. Sharing of interests and problems among workers in a department or institute is also valuable in promoting a stimulating atmosphere in which to work. Enthusiasm is infectious and is the best safeguard against the doldrums.

Conditioned thinking

Psychologists have observed that once we have made an error, as for example in adding up a column of figures, we have a tendency to repeat it again and again. This phenomenon is known as the persistent error. The same thing happens when we ponder over a problem; each time our thoughts take a certain course, the more likely is that course to be followed the next time. Associations form between the ideas in the chain of thoughts and become firmer each time they are used, until finally the connections are so well established that the chain is very difficult to break. Thinking becomes conditioned just as conditioned reflexes are formed. We may have enough data to arrive at a solution to the problem, but, once we have adopted an unprofitable line of thought, the oftener we pursue it, the harder it is for us to adopt the profitable line. As Nicolle says, "The longer you are in the presence of a difficulty, the less likely you are to solve it."

Thinking also becomes conditioned by learning from others by word of mouth or by reading. In the first chapter we discussed the adverse effect on originality of uncritical reading. Indeed, all learning is conditioning of the mind. Here, however, we are concerned with the effects of conditioning which are unprofitable for our immediate purpose, that of promoting original thought. This does not only concern learning or being conditioned to incorrect opinions for, as we have seen in the first chapter, reading, even the reading of what is true so far as it goes, may have an adverse effect on originality.

The two main ways of freeing our thinking from conditioning are temporary abandonment and discussion. If we abandon a problem for a few days or weeks and then return to it the old thought associations are partly forgotten or less strong and often we can then see it in a fresh light, and new ideas arise. The beneficial effect of temporary abandonment is well shown by laying aside for a few weeks a paper one has written. On coming back to it, flaws are apparent that escaped attention before, and fresh pertinent remarks may spring into the mind.

Discussion is a valuable aid in breaking away from sterile lines of thought that have become fixed. In explaining a problem to another person, and especially to someone not familiar with that field of science, it is necessary to clarify and amplify aspects of it that have been taken for granted and the familiar chain of thought cannot be followed. Not infrequently it happens that while one is making the explanation, a new thought occurs to one without the other person having said a word. The same may happen during the delivery of a lecture, for when the teacher explains something he "sees" it more clearly himself than he had before. The other person, by asking questions, even ill-informed ones, may make the narrator break the established chain, even if only to explain the futility of the suggestion, and this may result in him seeing a new approach to the problem or the connection between two or more observations or ideas that he had not noticed before. The effect that questioning has on the mind might be Likened to the stimulus given to a fire by poking; it disturbs the settled arrangement and brings about new combinations. In disturbing fixed lines of thought, discussion is perhaps more likely to be helpful when carried on with someone not familiar with your field of work, for near colleagues have many of the same thought habits as yourself The writing of a review of the problem may prove helpful in the same way as the giving of a lecture.

A further useful application of the conception of conditioned thinking is that when a problem has defied solution it is best to start again right from the beginning, and if possible with a new approach. For example, I worked unsuccessfully for several years trying to discover the micro-organism which causes foot-rot in sheep. I met with repeated frustrations but each time I started again along the same lines, namely, trying to select the causal organism by microscopy and then isolating it in culture. This method seemed the sensible one to follow and only when I had exhausted all possibilities and was forced to abandon it, did I think of a fundamentally different approach to the problem, namely, to try mixed cultures on various media until one was found which was capable of setting up the disease. Work along these lines soon led to the solution of the problem.

Summary

Productive thinking is started off by awareness of a difficulty. A suggested solution springs into the mind and is accepted or rejected. New combinations in our thoughts arise from rational associations, or from fancy or perhaps chance circumstances. The fertile mind tries a large number and variety of combinations.

The scientific thinker becomes accustomed to withholding judgment and remaining in doubt when the evidence is insufficient. Imagination only rarely leads one to a correct answer, and most of our ideas have to be discarded. Research workers ought not to be afraid of making mistakes provided they correct them in good time.

Curiosity atrophies after childhood unless it is transferred to an intellectual plane. The research worker is usually a person whose curiosity is turned toward seeking explanations for phenomena that are not understood.

Discussion is often helpful to productive thinking and informal daily discussion groups in research institutes are valuable.

Once we have contemplated a set of data, the mind tends to follow the same line of thought each time and therefore unprofitable lines of thought tend to be repeated. There are two aids to freeing our thought from this conditioning; to abandon the problem temporarily and to discuss it with another person, preferably someone not familiar with our work.

References

[1] . Dewey, J. *How We Think*. (Boston: D.C. Heath & Co., 1933). Permission to quote kindly granted by D.C. Heath & Co., Boston.
[2] Taylor, E.L. "The Present-day Organisation of Veterinary Research in Great Britain: Its Strength and Weaknesses." *Vet. Rec*, No. 60, 451 (1948).
[3] Harding, Rosamund E.M. *An Anatomy of Inspiration*. (Cambridge: W. Heffer & Sons Ltd., 1942). Cambridge. Permission to quote kindly granted by W. Heffer & Sons Ltd., Cambridge.
[4] Planck, Max. *Where is Science Going?* Trans, by James Murphy. (London: George Allen & Unwin Ltd., 1933). Permission to quote kindly granted by George Allen & Unwin Ltd., London.

[5] Kekulé, F.A., quoted by J.R. Baker (1942) from Schutz, G. *Ber. deut. chem. Ges.*, No. 23, 1265 (1890).
[6] Einstein, Albert. Preface in *Where is Science Going?* by Max Planck Trans, by James Murphy. (London: George Allen & Unwin Ltd., 1933). Permission to quote kindly granted by George Allen & Unwin Ltd., London.
[7] George, W.H. *The Scientist in Action. A Scientific Study of his Methods.* (London: Wilhams & Norgate Ltd., 1936). Permission to quote kindly granted by Williams & Norgate Ltd., London.
[8] Tyndall, J. *Faraday as a Discoverer.* (London: Longmans, Green & Co., 1868).
[9] Darwin, F. *Life and Letters of C. Darwin.* (London: John Murray, 1888).
[10] Bartlett, F. *British Medical Journal*, Vol. I, p. 835 (1947).
[11] Koenigsberger, L. *Hermann von Helmholtz.* Trans, by F.A. Welby. (Oxford: Clarendon Press, 1906). Permission to quote kindly granted by Clarendon Press, Oxford.
[12] Herter, C.A. Chapter entitled "Imagination and Idealism" in Ed. Pearce, R.M. *Medical Research and Education.* (New York: Science Press, 1913) p. 487.
[13] Bulloch, W. *History of Bacteriology.* (London: Oxford University Press, 1938).
[14] Robertson, T. Brailsford. *The Spirit of Research.* (Adelaide, Preece and Sons, 1931).

9

Science, Irrationality, & Innovation

S. James Gates, Jr.

S. James Gates, Jr. is a theoretical physicist specializing in string theory at the University of Maryland College Park. Gates is well known as one of the earliest workers in the field that would later become known as string theory first coming across ideas of supersymmetry during his graduate work at MIT in the 1970s. He later contributed to the first comprehensive volume on supersymmetry, "Superspace: or, one thousand and one lessons in supersymmetry." In this article for the Kean Review, he talks about how irrationality versus pure reason often has a huge impact on the creative process and popularity of theories in science.

While I use the word "science" in my title, in fact my essay is really only about innovation in the part of theoretical physics about which I have some intimate knowledge and well founded opinions. Facts are the necessary basis of opinions, and by concentrating on this small part of science that I know best, I hope to illustrate some points applicable to science in general [1].

In the following essay, I will attempt to do three things by way of commenting on what I have seen of the process of scientific innovation at work. The first task is to provide a working definition of science. In the second part of this essay, I try to explain my understanding of the boundary of theoretical science, the place where research innovation takes place. While established science is governed by the rigorous use of mathematics and logic, I believe the boundary of science is quite a different place. It is a bit untamed. Logic and rigor still count for a lot. However, surprisingly a

---

* Reprinted with permission of the author and journal from S.J. Gates, Jr. "Science,'Irrationality,' and Innovation", *The Kean Review*, Vol. 1, 93-105 Spring/Summer 2008

type of "irrationality" seems present as the fundamental enabling element that permits scientific knowledge to expand. In the final part of this essay, I turn my commentary on the efforts, as I have viewed them, to establish a mathematically complete paradigm for superstring/M-theory.

Presently science, through its application to technology, has been remaking the rules by which the wealth of nations will be determined in the future. The principal agencies of this are the Internet and World Wide Web. Resulting from new developments in science such as "nano-science" and "genomic-based science," we appear to be living at the dawning of an age where the technological application of science will potentially have the ability to remake the meaning of the word "human." There is both great promise and peril in this.

Science - An error-correcting meme

The evolutionary biologist Richard Dawkins defined the term "meme" to represent a unit of cultural knowledge that in many ways demonstrates behavior that is remarkably similar to a gene [2]. By this definition, science qualifies as a meme. However, it is a very special sort of meme that I wish to spend some time to define. I argue that science is an error-correcting meme. That is, science acts to correct errors in the collective human understanding of the physical reality of which it is a part and to expand the breadth of this understanding.

While the use of logical rationality is a hallmark of science, there is what I think of as an "irrational" basis for science. The mathematician Kurt Gödel (1906-1978) essentially showed the impossibility to formulate a completely axiomatic system. Simply put, making working assumptions as a basis for deriving subsequent statements is unavoidable, and these assumptions precede and ultimately lie outside of the system of rational scientific analysis. In order to begin the work of science, I believe we must accept two propositions. First, that there are attributes of the universe whose properties and dynamical evolution are governed by rational laws that are independent of the consciousness of the observer. Secondly, we must assume that human consciousness is capable of discerning these and able to represent them in some form of human language.

All I have to say assumes these fundamental articles of faith, fundamental propositions, that allow for science, as we know it, to exist. Scientific culture promotes a rationalist reality-based view of the objective universe. This view of our universe, our working hypothesis, has been the chief enabler of scientific and technological progress over the last few centuries.

The field of physics has been deeply influenced by many great intellectuals. Galileo Galilei (1564-1642) "pushed" a particular way to view the world by use of a very specialized language—mathematics. Mathematics had been used well prior to Galileo in descriptions of the physical world (a discovery we can ascribe to Pythagoras (582-502 BCE)), but Galileo's studies of motion, trajectories, and rates of change led to a fundamental increase in the mathematical basis of the physical sciences. Galileo described Natural Philosophy, the intellectual and immediate predecessor to physics, in the following way:

> *Natural Philosophy is written in the great book which ever lies before our eyes—I mean the universe but we cannot understand it if we do not first learn the language in which it is written. The book is written in the mathematical language and the symbols are triangles, circles, and other geometrical figures, without whose help it is impossible to comprehend a single word of it; one wanders in vain through a dark labyrinth* [3]

Following after Galileo, Isaac Newton (1643-1727) solved the problem of motion with the invention of "the Calculus." This proved a fundamental breakthrough in the development of physics and all of the physical sciences. It is also interesting to note that the mathematician Gottfried Leibniz (1646-1716) at about the same time made this same leap. Even more interesting is recent evidence calculus may have been created by Archimedes (287-212 BCE) but lost to humanity until the work of Leibniz and Newton.

Humanists may be surprised to think of mathematics as a language, but this is a part of life for physicists. Mathematics is, in fact, a very interesting and strange language with many properties in common with other languages. I tell nonscientists to think about mathematics as a language because this is the way scientists make use of it. I also have described mathematics as an organ of perception. By this I mean theoretical physicists are working to gain insight into structures that make up levels of existence to which we have no direct access. We achieve this first by the means of mathematics. We "see" these levels first with mathematics.

An illustration of this process can be seen in the ability to detect atoms. Using present day technology in the form of "atomic force microscopes," individual atoms can be directly imaged. This has only become possible within the last decade. Yet in a very real way, physicists have been "seeing" atoms for about one hundred years. We can trace this especially back to one of the great works of Einstein in 1905. That year he wrote "On the Movement of Small Particles Suspended in Stationary Liquids Required by the Molecular Kinetic Theory of Heat," published on July 18, 1905.

This work of Einstein firmly established the existence of atoms and their size. Thus, we physicists have "seen" atoms by way of their mathematical description in theories that explain observable behavior in our world. For us, mathematics really is another means for viewing the universe. I often tell young people mathematics is an extrasensory perception organ. Of course, I am not the first person to notice this very strange property of mathematics. Charles Darwin (1809-1882) once said, "Mathematics seems to endow one with something like a new sense." For one who uses mathematics this way, it often seems to function like a "third eye of science." [4]

Even for those of us who make use of the tool of mathematics in physics, or science in general, it has a lot of surprising properties. One of these properties I refer to as "telepathy." All languages have the power to convey ideas from one mind to another. To this extent, the telepathic power of languages is ubiquitous. However, the precision of the transmittal of concepts using mathematics is striking when compared to other forms of human communication.

A property of the mathematical language is that if an object or idea has a mathematical description, there is automatically a way to communicate it in a precise and detailed way to the minds of others. This is true even though the others have no previous experience with the new discovery and even if there is no substantial experiential commonality between the members communicating. The mathematically precise understanding of the discovery is accessible to anyone with a sufficiency of mathematical "numeracy" (the analog of literacy). This is one reason why it is so often seen in science fiction writing that humans first communicate with putative aliens via the use of mathematics. Humanity's experience with mathematics suggests that it is a universal language for rational discourse about physical reality. Mathematics has this peculiar telepathic nature to it because of the precision it enforces in its users. In a very real way we can know with much more precision than with other forms of human communication what another person is thinking. This precision does not exist in the media of aural, written, graphical, or visual representations outside of mathematics. In mathematics, we share a common platform for the exchange of concepts.

Often nonscientists appear subject to an illusion that science uncovers truths for our species. This is not the work of science. It reveals theories about the structure of the universe. Albert Einstein once said,

> *It is difficult to attach a precise meaning to the term scientific "truth." Thus, the meaning of the word "truth" varies according to whether we deal with a fact of experience, a mathematical proposition, or a scientific theory.* [5]

While other fields of human thought are said to present truths, science only reveals theories. The use of the word theory recognizes that any paradigmatic explanation of facts (i.e. scientific observation) is a proposition that can be proven false. Any claim made to being a part of science must surrender *ab initio* to this property, and it must in principle allow (by the action and reasoning of scientists) for the claim to be proven false. I sometimes say, "Science is not actually about finding the truth. It is about making our beliefs about our home—the universe—less false."

Scientists are aware that ours is a clever species. Due to our cleverness, science must take into account we are also clever enough to fool ourselves. Accordingly, built into the structure of science there are mechanisms for error correction. This is the role of what has been called the scientific method and the means by which we discern arguments, observations, and experiments that provide a basis for our system of beliefs. Modern science most directly gained this as a legacy from Galileo. Einstein said of this circumstance,

> *Propositions arrived at by purely logical means are completely empty as regards reality. Because Galileo saw this, and particularly because he drummed it into the scientific world, he is the father of modern physics—indeed of modern science altogether.* [6]

A corollary to such a system is that scientists must be willing, when presented with a preponderance of evidence, to abandon beliefs previously held as correct. In science, there can be no ultimate certainty about one's scientific beliefs. Part of the charge to each new generation of scientists is to check and recheck the "canon" of this system of belief that is its inheritance from previous generations.

Among systems of belief, science is almost unique in this embrace of fallibility and limitations on human ability and perception. Instead of a weakness, this is the source of the strength of science. It can be argued this unremitting dedication to the refinement of our understanding of the universe gives science, through its application in technology, more power to alter the quality of human life than perhaps any other system of belief. Certainly human history over the course of the last several centuries supports this.

Science lies at the intersection of several different and not completely overlapping regions. One of these is the realm of human imagination, and mathematics is part of this. A second realm consists of physical reality, and the final realm is a subset of this that corresponds to that part in which the technical competence of our species permits it to observe and manipulate portions of physical reality. Science, as we know it, can only exist in the region where these three completely overlap. Not all are static. It is clear that what we call technological progress means that the technical competence of our species is expanding. The part of the circle of physical reality that lies outside of our technical competence and mathematical imaginations constitutes the realm of profoundly unknown parts of physical reality.

It is possible to observe phenomena without possessing the requisite mathematical ideas to explain and give complete context to the observations. An example is the phenomenon of "high temperature" superconductivity, for which there exists no accepted scientific theory. Perhaps the opposite example to this is the part of physics known as superstring/M-theory. Here we have lots of mathematical imagination, but no observational basis of this set of ideas. It remains a piece of "prephysics," "protophysics," or "putative physics." Some of its detractors even say that it is "metaphysics," i.e., not physics.

Science - At the Irrational Boundary of Theoretical Physics

Knowledge (mathematical, scientific, and technological) is finite. It possesses a boundary beyond which we are blind. The only human facility by which we go beyond this boundary is imagination. A theoretical physicist is essentially a mathematician who has the ability to calculate something useful about our cosmos. We imagine new answers and solutions. We make them up! However, as scientists we are charged with taking this marvelous facility and seeking nature's confirmation that we are less incorrect than with our previous theories. Einstein's comment was that it was the sad fate of most theories to be shown incorrect shortly after their conception. For those not so roughly treated, at most nature says, "Maybe." Again and again we go to the laboratory to see if the new paradigm gains support. Many times at the blackboard, I have told students, "Physics is not about what I write on the board. It is about what happens in the laboratory."

Science is thus always in a state of "tentativity" (if I may introduce such a term), a state mostly static but with punctuating dynamic periods of changes in scientific theories about the universe. This culture must accept that its most cherished theories at some point in the future will likely be changed. Such shifts from one theory to another are dramatic and vividly illustrate Einstein's dictum: "Imagination is more important than knowledge." [7] Geology has shown us that life on our planet has repeatedly had to face the challenge of mass extinctions, often apparently caused by violent shifts in the earth's environment. During these extinctions, large numbers of species come to the end of their existence. Something similar occurs in theoretical physics.

The "Newtonian/Bohr-Schrödinger-Heisenberg Extinction" illustrates such a boundary. Prior to this "event" (lasting approximately from 1905-1926), it was possible to argue that many different ideas might hold the explanation for why the puzzle of atomic spectra occurred in nature. At the conclusion of this period of transition, one such theory emerged in the community of theoretical physicists as the consensus answer…quantum theory. Quantum theory is considerably more mathematically sophisticated than its ancestor paradigm, classical Newtonian physics. As in evolutionary biology, the extinction yields a concept that is more complicated than what went before. Quantum theory is not derivable from classical Newtonian theory; it is a daring leap of the imagination beyond it. This history suggests that in theoretical physics, the creation of a genuinely new rational paradigm is itself an irrational process.

While I argue that here we can see "irrationality" once again emerge in a surprising way in the scientific enterprise and that appearance has large implications for governing the rate and pace of innovation, there is in this history another important point to note. Scientific belief is the product of consensus within a community bound together by a common set of shared values. This reliance on consensus is both a strength and a hindrance. Unlike in a democracy, not all scientists are created equal. This means the progress of the community is particularly influenced by the role of individuals that appear in its history. This fact explains why it is often the case that the use of a few names seem at first capable of assignment as the reason scientific progress occurs. The reality is that scientific progress is often, far more than generally realized, the result of an enormous amount of information exchanged between disparate thinkers around the world.

The process of developing new scientific knowledge is a difficult task, as is all innovation. Some ideas are more fruitful to follow as working propositions than others. However, at any given time in the marketplace of ideas, it is not at all clear which in fact are most fruitful for future progress. This means that any researcher in this environment faces making a decision about which ideas to pursue. A difficulty lies in the fact that whatever choice is made involves a commitment usually of enormous amounts of resources (time, effort, etc.) on the basis of a severe lack of relevant information. In this situation, one strategy is to trust the intuition of other scientists with a record of past achievement. This confers enormous power to any scientist chosen to play this role. In an area of research that is in a state of flux, it is often the case that a conventional wisdom arises. On the other hand, this strategy also lessens the number of approaches that are either attempted or even proposed for how to make progress. Thus, this strategy also has the very great possibility of stifling innovation. The theoretical physics community is faced somewhat with a Hobbesian choice…maximize efficiency or maximize region of exploration. Maintaining a healthy balance is the challenge that is constant. In this circumstance, decisions are made based only on intuition, guesses, etc. In other words, I argue that once again "irrationality" becomes a factor in the operation of an entire scientific community.

This aspect of "irrationality" also throws light onto other curious and often overlooked factors that can emerge. One of these I believe can be characterized as "style." When I discuss this with other theoretical physicists, the first reaction can be one of puzzlement: "What is the meaning of style in theoretical physics or mathematics?" Sometimes an initial reaction is, "Certainly no such thing exists in mathematics."

Anyone, I believe, who reads primary source literature in mathematics over a sufficiently long period of time can quickly conclude that there was a very different style to the writing and thinking about mathematics just over one hundred years ago. Some might argue that we have come to a sort of end of history and have advanced to the point that style in mathematical thinking will undergo no additional evolution. In my opinion this is unlikely. One reason for this is the rapidly increasing capabilities of computers.

Thus far throughout the history of our species, there has been essentially a single route to the mastery of mathematics. One had to first master the symbolic language used to express mathematical ideas. This is so ubiquitous that it is easy to lose sight of the fact that the symbolic representation of mathematical ideas is separate from the ideas. To understand this point, I have a favorite analogy. There is another area of human accomplishment that contains both a powerful, even universally accepted intellectual construct as well as a powerful symbolic way of representing ideas. This is the example of music. Music exists through the widely accessible experience of listening, but it also exists in the form of written scores. The latter mode of transmitting musical ideas is limited only to people who read music. The reason for the wide access to music by listening is due to the existence of a myriad of instruments, from the human larynx to woodwinds, strings, percussion devices, etc., for playing it.

To my thinking, computers are on their way to becoming to mathematics what orchestral instruments are to music. The ultimate application of computers in mathematics may not be a matter of simply computation. I believe it is possible that with sufficient time, computers may become the engines for CEC, computer-enabled-conceptualization. Using my analogy, the computer could perhaps permit a genius to explore realms of mathematics even though that person has never mastered the standard symbolic representations of mathematics that have been our tradition for millennia. Should this development occur, it would greatly increase the percentage of humanity that engages in the study of mathematics. Just as musical geniuses who do not read scores exist, there may arise mathematical geniuses who do not master traditional mathematical symbology. Without the difficulties of the usual approach to expressing mathematical ideas, it seems most likely that at least a new patois for mathematics might develop.

I argue that if the mathematical language upon which so much of the structure of theoretical physics rests and has played a substantial role in shaping its culture can be subject to changes in style, then so too can the field itself. This in turn will likely have an implication for the ultimate success of investigations in theoretical physics. Style seems likely linked to the sorts of irrational choices that are made. If this is the case, then style has indeed an important implication. Different styles of music exist and enrich the engagement and interaction humanity has with it. We are certainly not impoverished by the fact that musicians pursue different styles of the art form. As a working physicist, I believe I can perceive different national styles in the endeavors of theoretical physicists. These are very subtle, but I believe they make a major contribution to the vitality of the field.

The allowance to believe that such differences in style might exist explained for me what was a long-standing puzzling statement I heard many years ago made by Nobel Laureate Abdus Salam [8]. He suspected that when a sufficient number of people of the African diaspora started to engage in theoretical physics, something like jazz would appear. It was fifteen or twenty years before I had the intimate knowledge of theoretical physics necessary to interpret this statement well enough to understand his meaning. I believe he was talking about the idea of style of which I have been discussing. A concomitant notion is that this is why diversity matters. To maximize the breadth of investigations, the greatest intellectual diversity would seem a prudent condition to support.

Superstring/M-Theory

In the ordinary course of progress in theoretical physics, the process whereby "irrationality" plays a role is kept in check by the ever-present demand that whatever mathematical structure is under discussion should have as a metric its success and capability to describe observational data. In my own area of research that touches on superstring/M-theory, there is currently a popular debate underway on this point.

While I personally do not agree with the essential thrusts of the two separate works by Smolin and Woit [9] (eds. note: both of these authors have written books severely criticizing string theory and its methodologies), I believe it is illustrative to examine a few examples of the actual struggle underway to complete the purely mathematical understanding of this still emerging topic. Note I have emphasized the word "mathematical." Even if a conceptually complete version of superstring/M-theory exists, does it provide an accurate mathematical description of the physics of our cosmos?

Due to the mathematically incomplete nature of the current development of superstring/M-theory, there exist enormous possibilities for putative results as part of the conventional wisdom. One of the most widely accepted parts of the conventional wisdom about superstring/M-theory is that it predicts hidden dimensions. Using this as a lens, I wish to spend some time discussing how nature may not possess such hidden dimensions and yet might still be described by some version of superstring/M-theory.

Prior to Einstein, the scientific establishment believed that light was similar to all other waves. As all other waves move in a medium, it was thought there must exist a medium in which light waves propagate. This medium was called the "luminiferous aether" or "ether" for short. All efforts to detect the ether have failed even to this day. Like the ether, the idea of quarks—for a period of the history of theoretical physics—was one that had no critical observational support. To this day, no direct isolation of quarks has been made. Yet today, quarks are an accepted part of scientific lore. So why was one set of ideas about the unseen accepted and the other rejected? The answer is, of course, that though direct isolation of quarks has never been achieved, the indirect evidence for their existence is overwhelming. So in the history of theoretical physics these examples show that sometimes ideas must be given a provisional acceptance in order to find evidence to support the idea. This example allows us to wonder if the putative extra hidden dimensions in superstring/M-theory are ether-like (to eventually disappear) or quark-like (to eventually be supported by indirect evidence).

In our universe, the mathematical description of photons, particles of light, is provided by a set of functions called "the 4-potential." Photons are related to measureable electric and magnetic forces. However and most remarkably, photons described by different 4-potentials can nonetheless lead to exactly the same measurable electric and magnetic forces! Two photons described by two such 4-potentials are called "gauge equivalent." However, since the two 4-potentials are really distinct, there must be some way to tell them apart. There is. Shortly after the description of such 4-potentials was discovered, physicists realized that there was a mathematical quantity given the name "the gauge parameter." In the nineteen thirties, a physicist named Nicholas Kemmer (1911-1998) found the surprising result that the gauge parameters act like angles that one measures with a protractor, even if there are no hidden dimensions present!

By the end of the nineteen eighties, a remarkably complete mathematical description of the physics required to describe the universe at the level of subatomic particles had been completed. It is called the Standard Model. It not only constitutes a mathematically complete paradigm, it is the most rigorously tested piece of scientific theory that has ever existed in the history of our species [10]. It literally agrees with thousands and even tens of thousands of experimental observations. It is an impressive achievement not only in terms of the sheer number of experiments with which it agrees, but also in terms of the precision of some of these tests. One such test is called the "gee minus two" experiment.

The type of structure containing only the usual three dimensions of space, one dimension of time and multiple numbers of Kemmer angles is the fundamental mathematical backbone of the highly successful Standard Model. Mathematicians call this structure a "fiber bundle." In much of the current discussion of superstring/M-theory, one finds the statement that the mathematical self-consistency of the constructs requires hidden dimensions. If this were strictly so, experiments ruling out extra dimensions would also rule out the possibility that superstring/M-theory can describe our universe. It turns out, however, that this point—due to the lack of a mathematically complete formulation of superstring/M-theory—possesses a loophole. This loophole, as far as I can tell, is logically consistent and yet it has not become part of the conventional wisdom.

In my own work during the late eighties, completed in conjunction with W. Siegel [11], evidence was presented that the approach of Kemmer could be directly implemented in superstring/M-theory. The evidence supported the idea that no hidden dimensions are required for the mathematical consistency of superstring/M-theory. Readers not intimately familiar with the topic might be tempted to regard this as a recent breakthrough. This is not so; our work was completed in 1989.

My work with Siegel suggested direct connections to Kemmer angles and fiber bundles are consistent with the idea of superstring/M-theory without hidden dimensions. Throughout the middle to late eighties, several groups of physicists tackled the idea that perhaps string theory could describe a four-dimensional approach such as the type known to occur as the mathematical foundation of the Standard Model. Using an approach called "free fermions," one group of theorists extensively investigated such possibilities. In a different approach another collaboration used a technique called the "covariant lattice" to reach the same result [12]. Thus there were three different approaches to the formulation of superstring/M-theory without the inclusion of hidden dimensions and my work with Siegel was not unique in this regard. The unique feature of our work was the direct connection to Kemmer angles. I should perhaps mention that the presence of Kemmer angles in these models does not rule out hidden dimensions [13]. In fact, the first work completed with Siegel provided a description of conventional ten-dimensional string theory.

Someone outside the field might be tempted to ask why if these results were found didn't more physicists pay attention to them? This is a question that has many answers. First, all superstring/M-theory constructions that only contain the usual three dimensions of space, one dimension of time, and multiple numbers of Kemmer angles are significantly more complicated mathematically than the ones that contain hidden dimensions. Although there was no mathematical reason not to explore these alternative four-dimensional versions of superstring/M-theory, the majority of the community felt these were "inelegant" and lacked of a certain sense of beauty. I have long held that Nature is the only arbiter on this matter or as Einstein said, "If you are out to describe the truth, leave elegance to the tailor." [14] This is an example of one of the irrational choices to which I alluded in the previous chapter. In the absence of compelling evidence otherwise, the consensus by which the community makes progress went against these constructions.

More recently, with the motivation provided by the observational data that our cosmos seems to be presently in a phase of accelerated expansion, a number of physicists investigating superstring/M-theory have begun to look anew at these "inelegant" constructions under the guise of what is called "the landscape." [15] In the absence of a complete mathematical paradigm, such phase transitions in the conventional wisdom of superstring/M-theory are all but guaranteed to occur.

It is my belief that there are hints to suggest superstring/M-theory can become "ordinary" theoretical physics. To see some of this, a reconsideration of the Standard Model is a good example. To build the Standard Model, there are choices that must be made in order for it to be consistent with observational results. In fact, the Standard Model is only one of an infinite collection of mathematically consistent constructions called quantum relativistic field theories or field theories for short. In order to use the field theory to describe our world, there are approximately twenty numbers that must be fixed to get the physics correct. In addition to these numbers, choices have also have to be made about the Kemmer angles as well as the number of basic constituents of matter. These constituents are called quarks and leptons. Among the twenty or so numbers to be fixed are their masses. Thus field theory constitutes what I have called a lathe for physics. One definition of a lathe describes it as "a machine capable of remarkable precision and versatility, but requiring a skilled and experienced operator for its success." A lathe confers upon its user the power to make choices. While it has been the conventional wisdom for most of the history of superstring/M-theory that the presence of such choices in this area are not to be viewed favorably, my work with Siegel showed evidence of precisely such properties. In these works, choices could be made about the number and properties of the Kemmer angles. Similarly, choices could be made about structures that seemed to control the number of constituents.

In 1989 during a series of lectures in Poland, I made the following remark

> *This present situation is extremely unsatisfactory and leaves enormous room for speculation on the behavior of string theories. For example, one such problem concerns "the four dimensional vacua" of strings theories. Presently, it is widely believed that a unique vacuum for heterotic string theory should be chosen by the dynamics. I like to think of this as the "Ptolemaic view" of string theory. The belief in a unique vacuum is akin to a belief in a unique place for the earth in earlier cosmology. The alternate "Copernican view" is more to my liking. After all, why should only one vacuum be consistent? In general relativity, the next most complicated gauge theory, the vacuum is not determined by the theory. Until a nonperturbative understanding of strings is achieved, this question will remain unanswerable.* [16]

My remark refers to the irrational choices my intuition led me to believe would be important for progress in the field. In particular, the attribute of an infinity of mathematically consistent solutions was an expression of my belief that the ultimate formulation of string theory would probably be like ordinary field theory. At the time of this statement, the number of vacua according to the conventional wisdom was expected to be small. Currently there has been an increasing number of voices espousing a much larger number of vacua. In some of the current string literature, one can find discussions of ten raised to the power of hundreds as the possible number of vacua [17]. While I cannot quite claim that an infinite number of vacua has become a widely accepted idea, at least we can see motion in this direction.

I believe there is hope for superstring/M-theory to become a normal theory. The Kemmer angles found in my work with Siegel come in two distinct types. One type determines the forces among matter. The other type controls the types of matter given a mathematical description. More recently, mathematical objects called "fluxes" are found to exist in these equations and seem available to describe masses as seen in the laboratory. In my lectures in Poland, I stated one other "irrational" idea:

> *It seems unlikely without a geometrical understanding of string field theory, that much of this mostly unimaginable "new physics" can be studied. A geometrical approach would be helpful to study this question.*

The idea here is that the truly fundamental issues surrounding superstring/M-theory are not likely to be resolved until a genuine paradigm shift of the type that was driven by the introduction of special and general relativity or that of quantum theory. The successful completion of the program of covariant string and superstring field theory seemed and still seems to me the best hope of completing the superstring/M-theory paradigm. This is likely to create a new type of equation (like the Schrödinger equation) capable of generating a truly new paradigm in the field of theoretical physics. To this day, there is a highly specialized little-researched area called string field theory, where, my belief in "irrationality" suggests, will be found the ultimate answers to many of the puzzles in the field. It should be noted that instead of "irrationality," perhaps the word "intuition" might be applicable. On the research boundary, the use of intuition by a scientist is extremely important in performing a benefit/cost analysis of whether to pursue an incomplete idea. To illustrate how this might work, I will conclude this section with one final illustrative tale from the boundary.

During the period of 1982-1984, I had the pleasure of guiding a young graduate student named Barton Zwiebach as he undertook the task of writing a Ph.D. thesis at Caltech. Around 1995, a new topic called "D-p-branes" began to appear in the discussions of superstring/M-theory. By this time, Barton was an untenured professor of physics at M.I.T. and concerned because in all of his work, he had never considered including calculations with "D-p-branes" in his consideration of string field theory. The source of his concern was obvious. Since these mathematical constructs had suddenly appeared, could they be important in efforts to construct the long sought string version of Schrödinger's equation? Although I had not carried out a detailed study of "D-p-branes," my irrational intuition told me they might be (and they have proved to be) useful for studying certain truncations of string theory. However, they did not seem likely to be of fundamental importance to string field theory itself. By 2004, following some brilliant deductions by a physicist named Sen, it was shown that "D-p-branes" did not seem at all important to superstring/M-theory owing to a property they possess called "supersymmetry." [18] In the end, all evidence pointed to "D-p-branes" not being relevant to string field theory nor required to achieve the long sought complete understanding of superstring/M-theory.

With this, I conclude my highly personalized view of a frontier of theoretical science that is still in an incomplete state. Superstring/M-theory, in spite of all the current criticism, is still in my opinion the best bet by far that our species has to make progress in understanding some of the ultimate questions of physics and our creation. One point about which much criticism has been written is that due to various reasons, superstring/M-theory has prevented promising alternatives from being more fully developed. As someone who attempts to keep an open mind about possibly promising approaches and as someone who has no special commitment solely to the idea of superstring/M-theory as the only possibility for providing progress, it has long been my sincere wish to see a competing set of ideas that seems as promising. As I hope my essay has shown, I have long been at variance with large amounts of what has become the conventional wisdom in the area. I have written about this and more importantly offered ideas and suggestions that are alternatives to those of the conventional wisdom. This I believe is the best that any committed theoretical physicist can do. One must offer concrete proposals as simple criticism alone offers no path for exploration.

Superstring/M-theory has an incredibly high "serendipity quotient," quite unlike any putative alternatives. It has shown a remarkable history of producing (yes only mathematical) coincidences that have traditionally been of the type associated with the first rank of mathematical constructions offering us the most profound views of creation. Rather than list many, there is only one with which I wish to close this essay.

The prime motive in the last thirty years of Einstein's research career was his drive to find what he named "the unified field theory." Einstein's theory of general relativity (GR) is, indeed, a theory about space and time, but it is almost a sterile theory in one way. It can be used in such a way as to be like the architectural plans of a magnificent but empty stage. In order for the play we call "physical reality" to unfold, it is necessary to fill that stage with actors. For practitioners of GR, this is done with the use of a device called "the stress-energy tensor." This mathematical construct is used to introduce electrons, photons, quarks, and related "beasts" into Einstein's equation. These are the actors in physical reality and the stuff of which we are made. But in GR, we are only an afterthought. If I may paraphrase Einstein, in the GR theory, God would have had a choice as to whether we should exist or not. These equations can describe a universe composed only of space and time that is mathematically consistent without us. In the confines of such a mathematical construct there arises a question: "Why should any matter, including us, exist at all?"

Superstring/M-theory of the four-dimensional variety in this way is fundamentally different. If the equations that describe our universe are not those given by GR but instead those given by, most especially, a four-dimensional superstring/M-theory, then the universe has no choice! It is a mathematical fact that no superstring capable of describing the three spatial dimensions and one temporal dimension is consistent without also necessarily describing quantities akin to our basic building blocks.

In superstring/M-theory, we are not an afterthought. As GR welded together space and time in an unprecedented way, superstring/M-theory welds together space and time as well as matter and energy in an unprecedented way. Superstring/M-theory says if there is a four-dimensional space-time stage for the performance called physical reality, then that stage must be filled with actors. From general relativity and quantum theory, we see that we are kin and kindred to the stars themselves. Superstring/M-theory says that each of us can trace our familial tree even further to space-time itself, the essence of physical reality. Thus, within the structure of superstring/M-theory, we are more intimately connected to the universe than it is possible to conceive using concepts from any other idea yet proposed.

Granted that at the end of the day, to become a normal theory, superstring/M-theory must confront some observational data about our cosmos. This is the ultimate arbitration in theoretical physics. For the theoretical physicists, unlike the musical composer, it is the audience of nature that counts, not the nature of the audience. Einstein, of course, was well aware of this. But he also was not divorced from philosophical arguments. I believe a philosophical point is also the most profound reason Einstein ardently pursued the quest for the "unified field theory." Using his language, Einstein wanted to know if God had a choice in how the universe was put together. Four-dimensional superstring/M-theory serendipitously answers, "Not as much as could have been," and all other claimed alternatives are silent.

For the reader who wishes to more deeply probe many of the aspects discussed in the final part of this essay, I have recently completed an accessible work aimed at the non-specialist where much detailed discussion can be found. I invite any interested party to pursue this and I hope my efforts in that direction are found to be efficacious [19].

Notes

[1] Portions of this essay derive from "On the Universality of Creativity in the Liberal Arts and Sciences," a paper delivered February 18, 2005 at Westmont College in Santa Barbara, California and published in *Beyond Two Cultures: The Sciences as Liberal Arts. The collected papers delivered at the Fifth Annual Conversation on the Liberal Arts.* Ed. Hoeckley , Christopher (Santa Barbara, CA: Gaede Institute for the Liberal Arts, 2005).

[2] Dawkins, Richard, *The Selfish Gene* (London: Oxford University Press, 1976)

[3] Galilei, Galileo, "The Assayer" in *Discoveries and Opinions of Galileo* translated by S. Drake (New York: Doubleday Anchor, 1957).

[4] *Charles Darwin in Mathematical Maxims and Minims* Ed. N. Rose (Raleigh, NC: Rome Press Inc., 1988).

[5] Einstein, Albert *Ideas and Opinion* (1954; reprint New York: Three Rivers Press Inc., 1982).

[6] Ibid.

[7] Einstein, Albert "What Life Means to Einstein," *The Saturday Evening Post*, October 26, 1929.

[8] Abdus Salam to S. James Gates, Jr. in private conversation, approximately 1982.

[9] Smolin, Lee *The Trouble with Physics: The Rise of String Theory, the Fall of a Science, and What Comes Next* (New York: Houghton Mifflin, 2006). Also Woit, Peter *Not Even Wrong: The Failure of String Theory and the Search for Unity in Physical Law* (New York: Basic Books, 2006).

[10] Oerter, Robert *The Theory of Almost Everything. The Standard Model, the Unsung Triumph of Modern Physics* (New York: Pi Press, 2005).

[11] Gates, S.J. and Siegel, W. *Physics Letters B* 2068 (1988): 631.

[12] Antoniadis, I., Bachas, C., Kounnas, C., and Windey, P. *Physics Letters* 1718 (1986): 51; Kawai, H., Lewellen, D., Tye, S.H.H. *Physics Letters B* 191 (1987): 63; *Nuclear Physics B* 288 (1987): 1. Lauer, J., Lust, D., Theisen, S. *Nuclear Physics B* 304 (1988): 236.

[13] Randall, L. *Warped Passages: Unraveling the Mysteries of the Universe's Hidden Dimensions* (New York: Ecco, 2005). Also Krauss, L. *Hiding in the Mirror: The Mysterious Allure of Extra Dimensions from Plato to String Theory and Beyond* (New York: Viking Adult, 2006).

[14] Einstein, Albert in *The Quotable Einstein*, Ed. Calaprice, A. (Princeton: Princeton University Press, 1996).
[15] Susskind, Leonard *The Cosmic Landscape: String Theory and the Illusion of Intelligent Design* (New York: Little Brown & Co., 2006).
[16] Gates, S. James, Jr. "Strings, Superstrings, and Two-Dimensional Lagrangian Field Theory," In *Functional Integration, Geometry, and Strings* Eds. Haba, Z. and Sobczyk, J. (Amsterdam: Birkhauser, 1989).
[17] Douglas, Michael, R. "Understanding the Landscape." http://arxiv.org/abs/hep-th/0602266.
[18] Sen, Ashoke, "Descent Relations Among Bosonic D-Branes," *International Journal of Modern Physics*, A14 (1999): 4061. http://www.arxiv.org/abs/hep-th/9902105.
[19] Gates, S. James, Jr. "Superstring Theory: The DNA of Reality," A DVD lecture collection available through The Teaching Company. http://www.teach12.com/teach12.asp?A1-16281.

10

The Art of Scientific Investigation: Chapter 7: Reason William Ian Beardmore Beveridge

"Discovery should come as an adventure rather than as the result of a logical process of thought. Sharp, prolonged thinking is necessary that we may keep on the chosen road, but it does not necessarily lead to discovery."
—Theobald Smith

Limitations and hazards

Before considering the role of reason in research it may be useful to discuss the limitations of reason. These are more serious than most people realise, because our conception of science has been given us by teachers and authors who have presented science in logical arrangement and that is seldom the way in which knowledge is actually acquired.

Everyday experience and history teach us that in the biological and medical sciences reason seldom can progress far from the facts without going astray. The scholasticism and authoritarianism prevailing during the Middle Ages was incompatible with science. With the Renaissance came a change in outlook: the belief that things ought and must behave according to accepted views (mostly taken from the classics) was supplanted by a desire to observe things as they really are, and human knowledge began to grow again. Francis Bacon had a great influence on the development of science mainly, I think, because he showed that most discoveries had been made empirically rather than by use of deductive logic. In 1605 he said:

* Reprinted from W.I.B. Beveridge, The Art of Scientific Investigation (New York: W.W. Norton, 1957) 82-95 under Creative Commons Attribution-ShareAlike License 3.0

Men are rather beholden...generally to chance, or anything else, than to logic, for the invention of arts and sciences, [1]

and in 1620,

the present system of logic rather assists in confirming and rendering inveterate the errors founded on vulgar notions, than in searching after truth, and is therefore more hurtful than useful. [2]

Later the French philosopher René Descartes made people realise that reason can land us in endless fallacies. His golden rule was: "Give unqualified assent to no propositions but those the truth of which is so clear and distinct that they cannot be doubted."

Every child, indeed one might even say, every young vertebrate, discovers gravity; and yet modern science with all its knowledge cannot yet satisfactorily "explain" it. Not only are reason and logic therefore insufficient to provide a means of discovering gravity without empirical knowledge of it, but all the reason and logic applied in classical times did not even enable intelligent men to deduce correctly the elementary facts concerning it.

F.C.S. Schiller, a modem philosopher, has made some illuminating comments on the use of logic in science and I shall quote from him at length:

Among the obstacles to scientific progress a high place must certainly be assigned to the analysis of scientific procedure which logic has provided....It has not tried to describe the methods by which the sciences have actually advanced, and to extract...rules which might be used to regulate scientific progress, but has freely re-arranged the actual procedure in accordance with its prejudices, for the order of discovery there has been substituted an order of proof. [3]

Credence of the logician's view has been encouraged by the method generally adopted in the writing of scientific papers. The logical presentation of results which is usually followed is hardly ever a chronological or full account of how the investigation was actually carried out, for such would often be dull and difficult to follow and, for ordinary purposes, wasteful of space. In his book on the writing of scientific papers, Allbutt specifically advocates that the course of the research should not be followed but that a deductive presentation should be adopted. To quote again from Schiller, who takes an extreme view:

> *It is not too much to say that the more deference men of science have paid to logic, the worse it has been for the scientific value of their reasoning...fortunately for the world, however, the great men of science have usually been kept in salutary ignorance of the logical tradition.*

He goes on to say that logic was developed to regulate debates in the Greek schools, assemblies and law-courts. It was necessary to determine which side won, and logic served this purpose, but it should not occasion surprise that it is quite unsuitable in science, for which it was never intended. Many logicians emphatically declare that logic, interested in correctness and validity, has nothing at all to do with productive thinking.

Schiller goes even further in his criticism of traditional logic and says that not only is it of little value in making new discoveries, but that history has shown it to be of little value in recognising their validity or ensuring their acceptance when they have been proclaimed. Indeed, logical reasoning has often prevented the acceptance of new truths, as is illustrated by the persecution to which the great discoverers have so often been subjected.

> *The slowness and difficulty with which the human race makes discoveries and its blindness to the most obvious facts, if it happens to be unprepared or unwilling to see them, should suffice to show that there is something gravely wrong about the logician's account of discovery.*

Schiller was protesting mainly against the view of the scientific method expounded by certain logicians in the latter half of the nineteenth century. Most modem philosophers concerning themselves with the scientific method do not interpret this phrase as including the art of discovery, which they consider to be outside their province. They are interested in the philosophical implications of science.

Wilfred Trotter [4] also had some provocative things to say about the poor record which reason has in the advancement of scientific knowledge. Not only has it few discoveries to its credit compared to empiricism, he says, but often reason has obstructed the advance of science owing to false doctrines based on it. In medicine particularly, practices founded on reason alone have often prevailed for years or centuries before someone with an independent mind questioned them and in many cases showed they were more harmful than beneficial.

Logicians distinguish between inductive reasoning (from particular instances to general principles, from facts to theories) and deductive reasoning (from the general to the particular, applying a theory to a particular case). In induction one starts from observed data and develops a generalisation which explains the relationships between the objects observed. On the other hand, in deductive reasoning one starts from some general law and applies it to a particular instance. Thus in deductive reasoning the derived conclusion is contained within the original premise, and should be true if the premise is true.

Since deduction consists of applying general principles to further instances, it cannot lead us to new generalisations and so cannot give rise to major advances in science. On the other hand the inductive process is at the same time less trustworthy but more productive. It is more productive because it is a means of arriving at new theories, but is less trustworthy because starting from a collection of facts we can often infer several possible theories, all of which cannot be true as some may be mutually incompatible; indeed none of them may be true.

In biology every phenomenon and circumstance is so complex and so poorly understood that premises are not clear-cut and hence reasoning is unreliable. Nature is often too subtle for our reasoning. In mathematics, physics and chemistry the basic premises are more firmly established and the attendant circumstances can be more rigidly defined and controlled. Therefore reason plays a rather more dominant part in extending knowledge in these sciences. Nevertheless the mathematician Poincare said:

Logic has very little to do with discovery or invention.

Similar views were expressed by Planck and Einstein. The point here is that inductions are usually arrived at not by the mechanical application of logic but by intuition, and the course of our thoughts is constantly guided by our personal judgment. On the other hand the logician is not concerned with the way the mind functions but with logical formulation.

From his experience in finding that his hypotheses always had to be abandoned or at least greatly modified Darwin learnt to distrust deductive reasoning in the biological sciences. He said:

I must begin with a good body of facts, and not from principle, in which I always suspect some fallacy. [5]

A basic difficulty in applying reason in research derives from the fact that terms often cannot be defined accurately and premises are seldom precise or unconditionally true. Especially in biology premises are only true under certain circumstances. For careful reasoning and clarity of thought one should first define the terms one uses but in biology exact definitions are often difficult or impossible to arrive at. Take, for example, the statement "influenza is caused by a virus." Influenza was originally a clinical concept, that is to say, a disease defined on clinical characters. We now know that diseases caused by several different microbes have been embraced by what the clinician regards as influenza. The virus worker would now prefer to define influenza as a disease caused by a virus with certain characters. But this only passes on the difficulty to the defining of an influenza virus which in turn escapes precise definition.

These difficulties are to some extent resolved if we accept the principle that in all our reasoning we can deal only in probabilities. Indeed much of our reasoning in biology is more aptly termed speculation.

I have mentioned some limitations inherent in the application of logical processes in science; another common source of error is incorrect reasoning, such as committing some logical fallacy. It is a delusion that the use of reason is easy and needs no training or special caution. In the following section I have tried to outline some general precautions which it may be helpful to keep in mind in using reason in research.

Some safeguards in use of reason in research

The first consideration is to examine the basis from which we start reasoning. This involves arriving at as clear an understanding as possible of what we mean by the terms we employ, and examining our premises. Some of the premises may be well-established facts or laws, while others may be purely suppositions. It is often necessary to admit provisionally some assumptions that are not well established, in which case one needs to be careful not to forget that they are only suppositions. Michael Faraday warned against the tendency of the mind "to rest on an assumption" and when it appears to fit in with other knowledge to forget that it has not been proved. It is generally agreed that unverified assumptions should be kept down to the bare minimum and the hypothesis with the fewest assumptions is to be preferred. (This is known as the maxim of parsimony, or "Occam's Razor." It was propounded by William of Occam in the fourteenth century.)

How easy it is for unverified assumptions to creep into our reasoning unnoticed! They are often introduced by expressions such as "obviously," "of course," "surely." I would have thought that it was a fairly safe assumption that well-fed animals live longer on the average that underfed ones, but in recent experiments mice whose diet was restricted to a point where their growth rate was below normal lived much longer than mice allowed to eat as much as they wished.

Having arrived at a clear understanding of the basis from which we start, at every step in our reasoning it is essential to pause and consider whether all conceivable alternatives have been taken into account. The degree of uncertainty or supposition is usually greatly magnified at each step.

It is important not to confuse facts with their interpretations—that is to say, to distinguish between data and generalisations. Facts are particular observational data relating to the past or present. To take an obvious illustration: it may be a fact that when a certain drug was administered to rabbits it killed them, but to say that the drug is poisonous for rabbits is not a statement of a fact but a generalisation or law arrived at by induction. The change from the past tense to the present usually involves stepping from the facts to the induction. It is a step which must often be taken but only with an understanding of what one is doing. Confusion may also arise from the way in which the results are interpreted: strictly the facts arising from experiments can only be described by a precise statement of what occurred. Often in describing an experiment we interpret the results into other terms, perhaps without realising we are departing from a statement of the facts.

A difficulty we are always up against is that we have to argue from past and present to the future. Science, to be of value, must predict. We have to reason from data obtained in the past by experiment and observation, and plan accordingly for the future. This presents special difficulties in biology because, owing to the incompleteness of our knowledge, we can seldom be sure that changed circumstances in the future may not influence the results.

Take, for example, the testing of a new vaccine against a disease. The vaccine may prove effective in several experiments but we must still be cautious in saying it will be effective in future. Influenza vaccine gave a considerable degree of protection in large scale trials in U.S.A. in 1943 and 1945, but against the next epidemic in 1947 it was of no value. Regarded as a problem in logic the position is that by inductive inference from our data we arrive at a generalisation (for instance, that the vaccine is effective). Then in future when we wish to guard against the disease we use this generalisation deductively and apply it to the particular practical problem of protecting certain people. The difficult point in the reasoning is, of course, making the induction. Logic has little to say here that is of help to us. All we can do is to refrain from generalising until we have collected fairly extensive data to provide a wide basis for the induction and regard as tentative any conclusion based on induction or, as we more often hear in everyday language, be cautious with generalisations. Statistics help us in drawing conclusions from our data by ensuring that our conclusions have a certain reliability, but even statistical conclusions are strictly valid only for events which have already occurred.

Generalisations can never be *proved.* They can be tested by seeing whether deductions made from them are in accord with experimental and observational facts, and if the results are not as predicted, the hypothesis or generalisation may be *disproved.* But a favourable result does not prove the generalisation, because the deduction made from it may be true without its being true. Deductions, themselves correct, may be made from palpably absurd generalisations. For instance, the truth of the hypothesis that plague is due to evil spirits is not established by the correctness of the deduction that you can avoid the disease by keeping out of the reach of the evil spirits. In strict logic a generalisation is never proved and remains on probation indefinitely, but if it survives all attempts at disproof it is accepted in practice, especially if it fits well into a wider theoretical scheme.

If scientific logic shows we must be cautious in arriving at generalisations ourselves, it shows for the same reasons that we should not place excessive trust in any generalisation, even widely accepted theories or laws. Newton did not regard the laws he formulated as the ultimate truth, but probably most following him did until Einstein showed how well-founded Newton's caution had been. In less fundamental matters how often do we see widely accepted notions superseded!

Therefore the scientist cannot afford to allow his mind to become fixed, with reference not only to his own opinions but also to prevailing ideas. Theobald Smith said:

> *Research is fundamentally a state of mind involving continual re-examination of doctrines and axioms upon which current thought and action are based. It is, therefore, critical of existing practices.* [6]

No accepted idea or "established principle" should be regarded as beyond being questioned if there is an observation challenging it. Bernard wrote:

> *If an idea presents itself to us, we must not reject it simply because it does not agree with the logical deductions of a reigning theory.*

Great discoveries have been made by means of experiments devised with complete disregard for well accepted beliefs. Evidently it was Darwin who introduced the expression "fool's experiment" to refer to such experiments, which he often undertook to test what most people would consider not worth testing.

People in most other walks of life can allow themselves the indulgence of fixed ideas and prejudices which make thinking so much easier, and for all of us it is a practical necessity to hold definite opinions on many issues in everyday life, but the research worker must try to keep his mind malleable and avoid holding set ideas in science. We have to strive to keep our mind receptive and to examine suggestions made by others fairly and on their own merits, seeking arguments for as well as against them. We must be critical, certainly, but beware lest ideas be rejected because an automatic reaction causes us to see only the arguments against them. We tend especially to resist ideas competing with our own.

A useful habit for scientists to develop is that of not trusting ideas based on reason only. As Trotter says, they come into the mind often with a disarming air of obviousness and certainty.

Some consider that there is no such thing as pure reasoning, that is to say, except where mathematical symbols are involved.

Practically all reasoning is influenced by feelings, prejudice and past experience, albeit often subconsciously. Trotter wrote:

> *The dispassionate intellect, the open mind, the unprejudiced observer, exist in an exact sense only in a sort of intellectualist folklore; states even approaching them cannot be reached without a moral and emotional effort most of us cannot or will not make.*

A trick of the mind well known to psychologists is to "rationalise," that is, to justify by reasoned argument a view which in reality is determined by preconceived judgment in the subconscious mind, the latter being governed by self-interest, emotional considerations, instinct, prejudice and similar factors which the person usually does not realise or admit even to himself In somewhat similar vein is W.H. George's warning against believing that things in nature ought to conform to certain patterns or standards and regarding all exceptions as abnormal. He says that the "should-ought mechanism" has no place whatever in research, and its complete abandonment is one of the foundation stones of science. It is premature, he considers, to worry about the technique of experimentation until a man has become dissatisfied with the "should-ought" way of thinking.

It has been said by some that scientists should train themselves to adopt a disinterested attitude to their work. I cannot agree with this view and think the investigator should try to exercise sufficient self-control to consider fairly the evidence against a certain outcome for which he fervently hopes, rather than to try to be disinterested. It is better to recognise and face the danger that our reasoning may be influenced by our wishes. Also it is unwise to deny ourselves the pleasure of associating ourselves wholeheartedly with our ideas, for to do so would be to undermine one of the chief incentives in science.

It is important to distinguish between interpolation and extrapolation. Interpolating means filling in a gap *between* established facts which form a series. When one draws a curve on a graph by connecting the points one interpolates. Extrapolating is going *beyond* a series of observations on the assumption that the same trend continues. Interpolation is considered permissible for most purposes provided one has a good series of data to work from, but extrapolation is much more hazardous. Apparently obvious extensions of our theories beyond the field in which they have been tested often lead us astray. The process of extrapolation is rather similar to implication and is useful in providing suggestions.

A useful aid in getting a clear understanding of a problem is to write a report on all the information available. This is helpful when one is starting on an investigation, when up against a difficulty, or when the investigation is nearing completion. Also at the beginning of an investigation it is useful to set out clearly the questions for which an answer is being sought. Stating the problem precisely sometimes takes one a long way toward the solution. The systematic arrangement of the data often discloses flaws in the reasoning, or alternative lines of thought which had been missed. Assumptions and conclusions at first accepted as "obvious" may even prove indefensible when set down clearly and examined critically. Some institutions make it a rule for all research workers to furnish a report quarterly on the work done, and work planned. This is useful not only for the director to keep in touch with developments but also to the workers themselves. Certain directors prefer verbal reports which they consider more useful in helping the research worker "get his ideas straight."

Careful and correct use of language is a powerful aid to straight thinking, for putting into words precisely what we mean necessitates getting our own minds quite clear on what we mean. It is with words that we do our reasoning, and writing is the expression of our thinking. Discipline and training in writing is probably the best training there is in reasoning. Allbutt has said that slovenly writing reflects slovenly thinking, and obscure writing usually confused thinking. The main aim in scientific reports is to be as clear and precise as possible and make each sentence mean exactly what it is intended to and be incapable of other interpretation. Words or phrases that do not have an exact meaning are to be avoided because once one has given a name to something, one immediately has a feeling that the position has been clarified, whereas often the contrary is true.

> *A verbal cloak of ignorance is a garment that often hinders progress.* [7]

The role of reason in research

Although discoveries originate more often from unexpected experimental results or observations, or from intuitions, than directly from logical thought, reason is the principle agent in most other aspects of research and the guide to most of our actions. It is the main tool in formulating hypotheses, in judging the correctness of ideas conjured up by imagination and intuition, in planning experiments and deciding what observations to make, in assessing the evidence and interpreting new facts, in making generalisations and finally in finding extensions and applications of a discovery.

The methods and functions of discovery and proof in research are as different as are those of a detective and of a judge in a court of law. While playing the part of the detective the investigator follows clues, but having captured his alleged fact, he turns judge and examines the case by means of logically arranged evidence. Both functions are equally essential but they are different.

It is in "factual" discoveries in biology that observation and chance—empiricism—plays such an important part. But facts obtained by observation or experiment usually only gain significance when we use reason to build them into the general body of knowledge. Darwin said:

> *Science consists in grouping facts so that general laws or conclusions may be drawn from them.* [5]

In research it is not sufficient to collect facts; by interpreting them, by seeing their significance and consequences we can often go much further. Walshe considers that just as important as making discoveries is what we make *of* our discoveries, or for that matter, of those of other people [8]. To help retain and use information our minds require a rationalised, logically consistent body of knowledge. Hughlings Jackson said that

> *We have multitudes of facts, but we require, as they accumulate, organisations of them into higher knowledge; we require generalisations and working hypotheses.*

The recognition of a new general principle is the consummation of scientific study.

Discoveries originating from so-called chance observations, from unexpected results in experiments or from intuitions are dramatic and arrest attention more than progress resulting from purely rational experimentation in which each step follows logically on the previous one so that the discovery only gradually unfolds. Therefore the latter, less spectacular process may be responsible for more advances than has been implied in the other chapters of this book. Moreover, as Zinsser said:

> *The preparatory accumulation of minor discoveries and of accurately observed details...is almost as important for the mobilisation of great forward drives as the periodic correlation of these disconnected observations into principles and laws by the vision of genius.* [9]

Often when one looks into the origin of a discovery one finds that it was a much more gradual process than one had imagined.

In nutritional research, the discovery of the existence of the various vitamins was in a number of instances empirical, but subsequent development of knowledge of them was rational. Usually in chemotherapy, after the initial empirical discovery opening up the field, rational experimentation has led to a series of improvements, as in the development of sulphathiazole, sulphamerazine, sulphaguanidine, etc., following on the discovery of the therapeutic value of sulphanilamide, the first compound of this type found to have bacteriostatic properties.

As described in the Appendix, Fleming followed up a chance observation to discover that the mould *Penicillium notatum* produced a substance that had bacteriostatic properties and was non-toxic. However, he did not pursue it sufficiently to develop a chemotherapeutic agent and the investigation was dropped.

During the latter quarter of the last century and first part of this there were literally dozens of reports of discoveries of antibacterial substances produced by bacteria and fungi [10]. Even penicillin itself was discovered before Fleming or Florey [11]. Quite a number of writers had not only suggested that these products might be useful therapeutically but had employed them and in some instances good results seem to have been obtained [10]. But all these empirical discoveries were of little consequence until Florey, by a deliberately planned, systematic attack on the problem, produced penicillin in a relatively pure and stable form and so was able to demonstrate its great clinical value.

Often the original discovery, like the crude ore from the mine, is of little value until it has been refined and fully developed. This latter process, less spectacular and largely rational, usually requires a different type of scientist and often a team. The role of reason in research is not so much in exploring the frontiers of knowledge as in developing the findings of the explorers.

A type of reasoning not yet mentioned is reasoning by analogy, which plays an important part in scientific thought. An analogy is a resemblance between the relationship of things, rather than between the things themselves. When one perceives that the relationship between *A* and *B* resembles the relationship between *X* and *Y* on one point, and one knows that *A* is related to 5 in various other ways, this suggests looking for similar relationships between *X* and *Y*. Analogy is very valuable in suggesting clues or hypotheses and in helping us comprehend phenomena and occurrences we cannot see. It is continually used in scientific thought and language but it is as well to keep in mind that analogy can often be quite misleading and of course can never prove anything.

Perhaps it is relevant to mention here that modem scientific philosophers try to avoid the notion of cause and effect. The current attitude is that scientific theories aim at describing associations between events without attempting to explain the relationship as being causal. The idea of cause, as implying an inherent necessity, raises philosophical difficulties and in theoretical physics the idea can be abandoned with advantage as there is then no longer the need to postulate a connection between the cause and effect. Thus, in this view, science confines itself to description—"how," not "why."

This outlook has been developed especially in relation to theoretical physics. In biology the concept of cause and effect is still used in practice, but when we speak of *the* cause of an event we are really over-simplifying a complex situation. Very many factors are involved in bringing about an event but in practice we commonly ignore or take for granted those that are always present or well-known and single out as *the* cause one factor which is unusual or which attracts our attention for a special reason. The cause of an outbreak of plague may be regarded by the bacteriologist as the microbe he finds in the blood of the victims, by the entomologist as the microbe-carrying fleas that spread the disease, by the epidemiologist as the rats that escaped from the ship and brought the infection into the port.

Summary

The origin of discoveries is beyond the reach of reason. The role of reason in research is not hitting on discoveries—either factual or theoretical—but verifying, interpreting and developing them and building a general theoretical scheme. Most biological "facts" and theories are only true under certain conditions and our knowledge is so incomplete that at best we can only reason on probabilities and possibilities.

References

[1] Bacon, Francis. *The Advancement of Learning* (1605).
[2] Bacon, Francis. *Novum Organum* (1620).
[3] Schiller, F.C.S. "Scientific Discovery and Logical Proof." In Ed. Singer, C. *Studies in the History and Method of Science*, (Oxford: Clarendon Press, 1917). Permission to quote kindly granted by Clarendon Press, Oxford.
[4] Trotter, W. *Collected Papers of Wilfred Trotter.* (London: Oxford University Press, 1941). Permission to quote kindly granted by Oxford University Press, London.
[5] Darwin, F. *Life and Letters of C. Darwin.* (London: John Murray, 1888)
[6] Smith, T. *American Journal of the Medical Sciences*, No. 178, 740 (1929).
[7] Topley, W.W.C & Wilson, G.S. *The Principles of Bacteriology and Immunity.* (London: Edward Arnold & Co., 1929).
[8] Walshe, F.M.R. "Some general considerations on higher or post-graduate medical studies." *British Medical Journal*, Sept. 2nd, p. 297 (1944).
[9] Zinsser, H. *As I Remember Him.* (London: Macmillan & Co., Ltd; Boston: Brown & Co.; Atlantic Monthly Press 1940). Permission to quote kindly granted by the publishers.
[10] Florey, H. *British Medical Bulletin*, Vol. 4, 248 (1946).
[11] Peters, J.T. *Acta Medica Scandinavica*, Vol. 126, 60 (1940).

Part III
The Discovery Process

11

The Art of Discovery: Making Discoveries by Rule
Wilhelm Ostwald

Wilhelm Ostwald (September 2, 1853 – April 4, 1932)—was a Baltic German chemist and is considered one of the founders of physical chemistry. Much of his research centered on chemical reaction rates and for this he won the 1909 Nobel Prize in Chemistry. He also invented a process for the mass production of nitric acid which became important with the mass production of ammonia made possible by the Haber-Bosch process. In this article for Scientific American in 1910, he talks about the art of discovery and in contrary to previous contributors, assigns much less importance to intuition and emphasizes process.

When Schiller sent the first draft of his "Buergschaft" ("Security") to Goethe for criticism, he wrote: "I am curious to know whether I have been fortunate enough to discover all of the important motives which can be found in this subject. See if anything else occurs to you. This is one of the cases in which one can proceed in a definite manner and almost make discoveries by rule." These last words show how Schiller was struck by the possibility of making discoveries in accordance with definite technical rules, although such a process suggested itself to him. The general view is still very similar to Schiller's, and discovery by rule, either in poetical or technical works, seems like a contradiction in terms. We are accustomed to regard discoveries as something which cannot be commanded, but depend upon favorable opportunity.

* Reprinted from Wilhelm Ostwald, "The Art of Discovery: Making Discoveries by Rule," *Scientific American Supplement* Vol. 70 No. 1807 123-124, August 20, 1910; public domain

This rather mystical view is opposed to the sober and sordid fact that discovery has already been organized extensively on a commercial basis. I shall not describe how Edison, after developing his great discoveries, was "capitalized" by a company with the expressed object not only of exploiting the discoveries already made, but also of making other discoveries of equal importance. In the great industries, the machine shops, the electro-technical establishments, and especially in the chemical factories, we find laboratories of discovery in regular operation. The cool, calculating business heads of these establishments evidently find that the great outlay involved in these laboratories is judiciously expended, for they would immediately strike out the appropriations for this purpose if they found the laboratories unprofitable.

It is possible to regard these laboratories as a means of systematically making use of the chances of discovery. Priestley, who at the end of the eighteenth century enriched chemistry with so many discoveries, likened his method to that of a huntsman who goes into the fields and forests, not knowing what he shall find, or, whether he shall find anything. It is a well-known fact, however, that hunting is now carried on in a more systematic manner. For the amusement of royal personages, in particular, it has been found possible to eliminate chance and to replace it by certainty. We are now treading a similar royal road to discovery (editor's note: note the reference to the 'royal road' analogy in the biographical sketch from Helmholtz). Instead of strolling through the field and relying upon chance, we have organized a regular drive, so that only a poor shot can fail to bring down the game.

This improvement in the art of hunting evidently consists in the replacement of the chance movements of the individual hunter, who can only cover a small part of the field, by a complete covering of the field with huntsmen and stalkers. In other words, no possibility of escape is left to the game. The modern art of invention and discovery is based upon the same principle. It covers the entire field of possibilities with a systematic drive so that no facts can escape discovery unless the huntsmen are poor shots. I will now leave metaphors and turn to fact. If I speak more of scientific than of technical discoveries it is because I have more exact knowledge of the former than of the latter. From personal experience in both fields, I am convinced that they may be regarded as essentially identical for our purpose. As an example, I select an early research of the celebrated botanist, W. Pfeffer, concerning the swarm-spores of certain algae. The male flowers of these plants produce spores which move spontaneously through the water and reach the female flowers with perfect certainty. Pfeffer asked himself whether these movements might not be caused by some substance produced by the female flowers. He expressed the sap from a number of female flowers, placed it in a glass tube, and found that the swarm-spores directed their course as eagerly and accurately to the glass tube as to the female flowers. Thus, his question was answered in the affirmative, but now arose a second question-what substance causes this effect? The prospect of answering this question by direct chemical analysis of the flowers was hopeless, as the flowers contained hundreds of diverse organic compounds which the ablest chemist could not have isolated and identified. It was, therefore, necessary to attack the problem from the other side and to study the attraction exerted upon the swarm-spores by known substances; but this would have required many thousands of experiments with as many known organic compounds, and would have occupied a very long time. Pfeffer, therefore, treated the problem comprehensively by simply mixing together all of the substances on the top shelf of his cabinet and testing the effects of the mixture upon the swarm-spores. He proceeded in the same manner with the contents of the other shelves, until he found a mixture which attracted the spores. If we assume that there were 100 substances in this mixture, he had found that the attraction was due to one of these 100 substances.

Pfeffer next divided the whole number into the fifty on the right half of the shelf and the fifty on the left half, and thus limited the field to fifty substances by means of only two experiments. The group of fifty was similarly subdivided, again and again, until the substance which attracts the swarm-spores was isolated. It proved to be malic acid.

This is the whole secret of the art of discovery. The entire field of possibilities is divided into sections which can be controlled by the means at our command and each section is separately examined. By this method the particular part of the field which contains the solution of the problem cannot escape discovery.

It may possibly be objected that it is necessary to have a thorough knowledge of the subject in order to carry out this systematic subdivision of the field of possibilities, but the example quoted above shows that any method of subdivision which covers the whole field may be used. What is more superficial than the classification of chemical substances according to the shelves on which they happen to stand in a cabinet? Yet this apparently ludicrous method resulted in the solution of a very subtle problem.

If we prefer, we ran express the problem in learned language. Let us designate by *a*, *b*, *c*, *d*, etc. the various factors or circumstances which are concerned in the phenomenon under investigation. If we designate this phenomenon by *E*, the relation between it and its factors can be expressed by the general equation of $E=f(a, b, c, d, \text{etc.})$, which expresses the fact that the phenomenon *E* is a function of *a*, *b*, *c*, *d*, etc. In order to discover the mode of action of these factors, the most certain method consists in varying one or a number of them and leaving the others unchanged. The observed changes in the phenomenon must depend upon the factors which have been changed, and in this way their method of action can be studied. When a has been eliminated as ineffective the process is repeated with b and so on, until the effective factors are found. Then the entire phenomenon is within our grasp, and it becomes possible to plan a successful research. I am quite certain that the objection will be made that no inventions or discoveries are made in this mechanical fashion, but that genius unconsciously and instantly grasps the correct solution. This belief is only a tradition, and a very mischievous one. In every case in which we have personal descriptions of the labors of great men, we find that these men have worked as strenuously as any of us common mortals and with far greater devotion. The four notes which open Beethoven's Fifth Symphony, representing Fate knocking at the door, were developed step by step to their imposing simplicity, as we learn from the master's sketch books. Every master, in every field of endeavor, has become a master, not by stopping work earlier than others, but by keeping at it longer, and finding the possibility of improvement where others would have been content with the result already accomplished.

This tradition is mischievous because it induces the novice to rely upon fortunate chance. I have had abundant opportunity of observing this infantile disease in young investigators whom I have guided in their first steps in the path of discovery. When the systematic method, by which the problem could be gradually driven in and grasped, had been explained to them, they could seldom resist the temptation or seizing instantly upon some one of the countless possibilities, with the secret hope that they had "instinctively" found the right solution; for every enthusiastic beginner entertains the modest hope of being a genius. The uniform result was disillusionment and waste of time, for the systematic method not only ensures success, but is also, on the whole, the most economical of time and energy, as anyone familiar with the theory of probabilities can easily calculate.

There is, it is true, a scientific instinct, i.e., an unconscious trend of thought which leads to the selection, from many possibilities, of one suited to the purpose; but, as biologists regard every instinct as the result of a long process of natural selection, so the scientific instinct is developed, from long experience, in the latter part of the investigator's career. Then he can greatly shorten the process, but not without incurring the danger of one-sidedness.

Is it possible, then, for every man to become a successful discoverer by following the rules? The incredulous reader asks. No; no more than it is possible for every man to become a good violinist, or an expert mechanician. In order that this plan can be followed with success a sufficient endowment of imagination and of positive knowledge must be present.

The former facilitates the planning of the hunt the latter does the work of the beaters and drives the game from its hiding places. But although it is not possible for everyone to master the art of discovery, the art can still be learned. I have been convinced of this, to my consternation, in my own household. I am accustomed, at dinner, to submit to my boys various little technical problems, asking them to see what they can make of them. The boys learned the art of discovery so quickly, that at times I have been fairly overwhelmed by their achievements.

The art of discovery resembles all other arts and accomplishments. At first the prerogative of a few independent minds, the arts subsequently were acquired by pupils and imitators, although at first in a very imperfect manner. Then they gradually became common property, until finally some of them, like reading and writing, became a part of the intellectual inventory of every one. We have seen this development with our own eyes in the case of bicycling, and we are approaching a similar phase in the arts of discovery and invention. But although the general state of culture exhibits this progressive improvement, there will always be differences in the readiness with which individuals are able to utilize the common possession. On the other hand, it is in the nature of all such developments to diminish these differences, as the history of civilization abundantly proves.

12

Hypotheses In Physics
Henri Poincaré

Henri Poincaré (April 29, 1854 – July 17, 1912)—was a French mathematician and is widely considered one of the greatest mathematicians of the 20th century. His impact has reached far and wide across many disciplines but he is probably best known for his work on the stability of three body gravitational systems that preceded the discovery of chaotic dynamics and for being one of the first scientists to realize the necessity and importance of Einstein's special relativity, having laid part of the ground work for it by proving the invariance of Maxwell's Relations and thus the invariance of the speed of light in all reference frames. A believer in the importance of intuition for discovery, his writings in the next two chapters expound on his philosophy of science and creativity.

The Role Of Experiment And Generalization

Experiment is the sole source of truth. It alone can teach us anything new; it alone can give us certainty. These are two points that can not be questioned.

But then, if experiment is everything, what place will remain for mathematical physics? What has experimental physics to do with such an aid, one which seems useless and perhaps even dangerous?

And yet mathematical physics exists, and has done unquestionable service. We have here a fact that must be explained. The explanation is that merely to observe is not enough. We must use our observations, and to do that we must generalize. This is what men always have done; only as the memory of past errors has made them more and more careful, they have observed more and more, and generalized less and less.

---

* Reprinted from Henri Poincaré,"Hypotheses in Physics" in *The Foundations of Science* Translated by George Bruce Halstead (New York: The Science Press, 1921) 127-139; public domain.

Every age has ridiculed the one before it, and accused it of having generalized too quickly and too naively. Descartes pitied the lonians; Descartes, in his turn, makes us smile. No doubt our children will some day laugh at us.

But can we not then pass over immediately to the goal? Is not this the means of escaping the ridicule that we foresee? Can we not be content with just the bare experiment?

No, that is impossible; it would be to mistake utterly the true nature of science. The scientist must set in order. Science is built up with facts, as a house is with stones. But a collection of facts is no more a science than a heap of stones is a house.

And above all the scientist must foresee. Carlyle has somewhere said something like this: "Nothing but facts are of importance. John Lackland passed by here. Here is something that is admirable. Here is a reality for which I would give all the theories in the world." Carlyle was a fellow countryman of Bacon; but Bacon would not have said that. That is the language of the historian. The physicist would say rather: "John Lackland passed by here; that makes no difference to me, for he never will pass this way again."

We all know that there are good experiments and poor ones. The latter will accumulate in vain; though one may have made a hundred or a thousand, a single piece of work by a true master, by a Pasteur, for example, will suffice to tumble them into oblivion. Bacon would have well understood this; it is he who invented the phrase *Experimentum crucis*. But Carlyle would not have understood it. A fact is a fact. A pupil has read a certain number on his thermometer; he has taken no precaution; no matter, he has read it, and if it is only the fact that counts, here is a reality of the same rank as the peregrinations of King John Lackland. Why is the fact that this pupil has made this reading of no interest, while the fact that a skilled physicist had made another reading might be on the contrary very important? It is because from the first reading we could not infer anything. What then is a good experiment? It is that which informs us of something besides an isolated fact; it is that which enables us to foresee, that is, that which enables us to generalize.

For without generalization foreknowledge is impossible. The circumstances under which one has worked will never reproduce themselves all at once. The observed action then will never recur; the only thing that can be affirmed is that under analogous circumstances an analogous action will be produced. In order to foresee, then, it is necessary to invoke at least analogy, that is to say, already then to generalize.

No matter how timid one may be, still it is necessary to interpolate. Experiment gives us only a certain number of isolated points. We must unite these by a continuous line. This is a veritable generalization. But we do more; the curve that we shall trace will pass between the observed points and near these points; it will not pass through these points themselves. Thus one does not restrict himself to generalizing the experiments, but corrects them; and the physicist who should try to abstain from these corrections and really be content with the bare experiment, would be forced to enunciate some very strange laws.

The bare facts, then, would not be enough for us; and that is why we must have science ordered, or rather organized. It is often said experiments must be made without a preconceived idea. That is impossible. Not only would it make all experiment barren, but that would be attempted which could not be done. Every one carries in his mind his own conception of the world, of which he can not so easily rid himself. We must, for instance, use language; and our language is made up only of preconceived ideas and can not be otherwise. Only these are unconscious preconceived ideas, a thousand times more dangerous than the others.

Shall we say that if we introduce others, of which we are fully conscious, we shall only aggravate the evil? I think not. I believe rather that they will serve as counterbalances to each other—I was going to say as antidotes; they will in general accord ill with one another—they will come into conflict with one another, and thereby force us to regard things under different aspects. This is enough to emancipate us. He is no longer a slave who can choose his master.

Thus, thanks to generalization, each fact observed enables us to foresee a great many others; only we must not forget that the first alone is certain, that all others are merely probable. No matter how solidly founded a prediction may appear to us, we are never *absolutely* sure that experiment will not contradict it, if we undertake to verify it. The probability, however, is often so great that practically we may be content with it. It is far better to foresee even without certainty than not to foresee at all.

One must, then, never disdain to make a verification when opportunity offers. But all experiment is long and difficult; the workers are few; and the number of facts that we need to foresee is immense. Compared with this mass the number of direct verifications that we can make will never be anything but a negligible quantity.

Of this few that we can directly attain, we must make the best use; it is very necessary to get from every experiment the greatest possible number of predictions, and with the highest possible degree of probability. The problem is, so to speak, to increase the yield of the scientific machine.

Let us compare science to a library that ought to grow continually. The librarian has at his disposal for his purchases only insufficient funds. He ought to make an effort not to waste them. It is experimental physics that is entrusted with the purchases. It alone, then, can enrich the library. As for mathematical physics, its task will be to make out the catalogue. If the catalogue is well made, the library will not be any richer, but the reader will be helped to use its riches.

And even by showing the librarian the gaps in his collections, it will enable him to make a judicious use of his funds; which is all the more important because these funds are entirely inadequate. Such, then, is the role of mathematical physics. It must direct generalization in such a manner as to increase what I just now, called the yield of science. By what means it can arrive at this, and how it can do it without danger, is what remains for us to investigate.

The Unity Of Nature

Let us notice, first of all, that every generalization implies in some measure the belief in the unity and simplicity of nature. As to the unity there can be no difficulty. If the different parts of the universe were not like the members of one body, they would not' act on one another, they would know nothing of one another; and we in particular would know only one of these parts. We do not ask, then, if nature is one, but how it is one.

As for the second point, that is not such an easy matter. It is not certain that nature is simple. Can we without danger act as if it were? There was a time when the simplicity of Mariotte's law was an argument invoked in favor of its accuracy; when Fresnel himself, after having said in a conversation with Laplace that nature was not concerned about analytical difficulties, felt himself obliged to make explanations, in order not to strike too hard at prevailing opinion.

Today ideas have greatly changed; and yet, those who do not believe that natural laws have to be simple, are still often obliged to act as if they did. They could not entirely avoid this necessity without making impossible all generalization, and consequently all science. It is clear that any fact can be generalized in an infinity of ways, and it is a question of choice. The choice can be guided only by considerations of simplicity. Let us take the most commonplace case, that of interpolation. We pass a continuous line, as regular as possible, between the points given by observation. Why do we avoid points making angles and too abrupt turns? Why do we not make our curve describe the most capricious zigzags? It is because we know beforehand, or believe we know, that the law to be expressed can not be so complicated as all that.

We may calculate the mass of Jupiter from either the movements of its satellites, or the perturbations of the major planets, or those of the minor planets. If we take the averages of the determinations obtained by these three methods, we find three numbers very close together, but different. We might interpret this result by supposing that the coefficient of gravitation is not the same in the three cases. The observations would certainly be much better represented. Why do we reject this interpretation? Not because it is absurd, but because it is needlessly complicated. We shall only accept it when we are forced to, and that is not yet.

To sum up, ordinarily every law is held to be simple till the contrary is proved. This custom is imposed upon physicists by the causes that I have just explained. But how shall we justify it in the presence of discoveries that show us every day new details that are richer and more complex? How shall we even reconcile it with the belief in the unity of nature? For if everything depends on everything, relationships where so many diverse factors enter can no longer be simple.

If we study the history of science, we see happen two inverse phenomena, so to speak. Sometimes simplicity hides under complex appearances; sometimes it is the simplicity which is apparent, and which disguises extremely complicated realities. What is more complicated than the confused movements of the planets? What simpler than Newton's law? Here nature, making sport, as Fresnel said, of analytical difficulties, employs only simple means, and by combining them produces I know not what inextricable tangle. Here it is the hidden simplicity which must be discovered.

Examples of the opposite abound. In the kinetic theory of gases, one deals with molecules moving with great velocities, whose paths, altered by incessant collisions, have the most capricious forms and traverse space in every direction. The observable result is Mariotte's simple law. Every individual fact was complicated. The law of great numbers has reestablished simplicity in the average. Here the simplicity is merely apparent, and only the coarseness of our senses prevents our perceiving the complexity.

Many phenomena obey a law of proportionality. But why? Because in these phenomena there is something very small. The simple law observed, then, is only a result of the general analytical rule that the infinitely small increment of a function is proportional to the increment of the variable. As in reality our increments are not infinitely small, but very small, the law of proportionality is only approximate, and the simplicity is only apparent. What I have just said applies to the rule of the superposition of small motions, the use of which is so fruitful, and which is the basis of optics.

And Newton's law itself? Its simplicity, so long undetected, is perhaps only apparent. Who knows whether it is not due to some complicated mechanism, to the impact of some subtle matter animated by irregular movements, and whether it has not become simple only through the action of averages and of great numbers? In any case, it is difficult not to suppose that the true law contains complementary terms, which would become sensible at small distances. If in astronomy they are negligible as modifying Newton's law, and if the law thus regains its simplicity, it would be only because of the immensity of celestial distances.

No doubt, if our means of investigation should become more and more penetrating, we should discover the simple under the complex, then the complex under the simple, then again the simple under the complex, and so on, without our being able to foresee what will be the last term.

We must stop somewhere, and that science may be possible, we must stop when we have found simplicity. This is the only ground on which we can rear the edifice of our generalizations. But this simplicity being only apparent, will the ground be firm enough? This is what must be investigated.

For that purpose, let us see what part is played in our generalizations by the belief in simplicity. We have verified a simple law in a good many particular cases; we refuse to admit that this agreement, so often repeated, is simply the result of chance, and conclude that the law must be true in the general case.

Kepler notices that a planet's positions, as observed by Tycho, are all on one ellipse. Never for a moment does he have the thought that by a strange play of chance Tycho never observed the heavens except at a moment when the real orbit of the planet happened to cut this ellipse.

What does it matter then whether the simplicity be real, or whether it covers a complex reality? Whether it is due to the influence of great numbers, which levels down individual differences, or to the greatness or smallness of certain quantities, which allows us to neglect certain terms, in no case is it due to chance. This simplicity, real or apparent, always has a cause. We can always follow, then, the same course of reasoning, and if a simple law has been observed in several particular cases, we can legitimately suppose that it will still be true in analogous cases. To refuse to do this would be to attribute to chance an inadmissible role.

There is, however, a difference. If the simplicity were real and essential, it would resist the increasing precision of our means of measure. If then we believe nature to be essentially simple, we must, from a simplicity that is approximate, infer a simplicity that is rigorous. This is what was done formerly; and this is what we no longer have a right to do.

The simplicity of Kepler's laws, for example, is only apparent. That does not prevent their being applicable, very nearly, to all systems analogous to the solar system; but it does prevent their being rigorously exact.

The Role Of Hypothesis

All generalization is a hypothesis. Hypothesis, then, has a necessary role that no one has ever contested. Only, it ought always, as soon as possible and as often as possible, to be subjected to verification. And, of course, if it does not stand this test, it ought to be abandoned without reserve. This is what we generally do, but sometimes with rather an ill humor.

Well, even this ill humor is not justified. The physicist who has just renounced one of his hypotheses ought, on the contrary, to be full of joy; for he has found an unexpected opportunity for discovery. His hypothesis, I imagine, had not been adopted without consideration; it took account of all the known factors that it seemed could enter into the phenomenon. If the test does not support it, it is because there is something unexpected and extraordinary; and because there is going to be something found that is unknown and new.

Has the discarded hypothesis, then, been barren? Par from that, it may be said it has rendered more service than a true, hypothesis. Not only has it been the occasion of the decisive experiment, but, without having made the hypothesis, the experiment would have been made by chance, so that nothing would have been derived from it. One would have seen nothing extraordinary; only one fact the more would have been catalogued without deducing from it the least consequence.

Now on what condition is the use of hypothesis without danger? The firm determination to submit to experiment is not enough; there are still dangerous hypotheses; first, and above all, those which are tacit and unconscious. Since we make them without knowing it, we are powerless to abandon them. Here again, then, is a service that mathematical physics can render us. By the precision that is characteristic of it, it compels us to formulate all the hypotheses that we should make without it, but unconsciously.

Let us notice besides that it is important not to multiply hypotheses beyond measure, and to make them only one after the other. If we construct a theory based on a number of hypotheses, and if experiment condemns it, which of our premises is it necessary to change? It will be impossible to know. And inversely, if the experiment succeeds, shall we believe that we have demonstrated all the hypotheses at once? Shall we believe that with one single equation we have determined several unknowns?

We must equally take care to distinguish between the different kinds of hypotheses. There are first those which are perfectly natural and from which one can scarcely escape. It is difficult not to suppose that the influence of bodies very remote is quite negligible, that small movements follow a linear law, that the effect is a continuous function of its cause. I will say as much of the conditions imposed by symmetry. All these hypotheses form, as it were, the common basis of all the theories of mathematical physics. They are the last that ought to be abandoned.

There is a second class of hypotheses, that I shall term neutral. In most questions the analyst assumes at the beginning of his calculations either that matter is continuous or, on the contrary, that it is formed of atoms. He might have made the opposite assumption without changing his results. He would only have had more trouble to obtain them; that is all. If, then, experiment confirms his conclusions, will he think that he has demonstrated, for instance, the real existence of atoms?

In optical theories two vectors are introduced, of which one is regarded as a velocity, the other as a vortex. Here again is a neutral hypothesis, since the same conclusions would have been reached by taking precisely the opposite. The success of the experiment, then, can not prove that the first vector is indeed a velocity; it can only prove one thing, that it is a vector. This is the only hypothesis that has really been introduced in the premises. In order to give it that concrete appearance which the weakness of our minds requires, it has been necessary to consider it either as a velocity or as a vortex, in the same way that it has been necessary to represent it by a letter, either x or y. The result, however, whatever it may be, will not prove that it was right or wrong to regard it as a velocity any more than it will prove that it was right or wrong to call it x and not y.

These neutral hypotheses are never dangerous, if only their character is not misunderstood. They may be useful, either as devices for computation, or to aid our understanding by concrete images, to fix our ideas as the saying is. There is, then, no occasion to exclude them. The hypotheses of the third class are the real generalizations. They are the ones that experiment must confirm or invalidate. Whether verified or condemned, they will always be fruitful. But for the reasons that I have set forth, they will only be fruitful if they are not too numerous.

Origin Of Mathematical Physics

Let us penetrate further, and study more closely the conditions that have permitted the development of mathematical physics. We observe at once that the efforts of scientists have always aimed to resolve the complex phenomenon directly given by experiment into a very large number of elementary phenomena.

This is done in three different ways: first, in time. Instead of embracing in its entirety the progressive development of a phenomenon, the aim is simply to connect each instant with the instant immediately preceding it. It is admitted that the actual state of the world depends only on the immediate past, without being directly influenced, so to speak, by the memory of a distant past. Thanks to this postulate, instead of studying directly the whole succession of phenomena, it is possible to confine ourselves to writing its 'differential equation.' For Kepler's laws we substitute Newton's law. Next we try to analyze the phenomenon in space. What experiment gives us is a confused mass of facts presented on a stage of considerable extent. We must try to discover the elementary phenomenon, which will be, on the contrary, localized in a very small region of space.

Some examples will perhaps make my thought better understood. If we wished to study in all its complexity the distribution of temperature in a cooling solid, we should never succeed. Everything becomes simple if we reflect that one point of the solid can not give up its heat directly to a distant point; it will give up its heat only to the points in the immediate neighborhood, and it is by degrees that the flow of heat can reach other parts of the solid. The elementary phenomenon is the exchange of heat between two contiguous points. It is strictly localized, and is relatively simple, if we admit, as is natural, that it is not influenced by the temperature of molecules whose distance is sensible. I bend a rod. It is going to take a very complicated form, the direct study of which would be impossible. But I shall be able, however, to attack it, if I observe that its flexure is a result only of the deformation of the very small elements of the rod, and that the deformation of each of these elements depends only on the forces that are directly applied to it, and not at all on those which may act on the other elements.

In all these examples, which I might easily multiply, we admit that there is no action at a distance, or at least at a great distance. This is a hypothesis. It is not always true, as the law of gravitation shows us. It must, then, be submitted to verification. If it is confirmed, even approximately, it is precious, for it will enable us to make mathematical physics, at least by successive approximations.

If it does not stand the test, we must look for something else analogous; for there are still other means of arriving at the elementary phenomenon. If several bodies act simultaneously, it may happen that their actions are independent and are simply added to one another, either as vectors or as scalars. The elementary phenomenon is then the action of an isolated body. Or again, we have to deal with small movements, or more generally with small variations, which obey the well-known law of superposition. The observed movement will then be decomposed into simple movements, for example, sound into its harmonics, white light into its monochromatic components.

When we have discovered in what direction it is advisable to look for the elementary phenomenon, by what means can we reach it? First of all, it will often happen that in order to detect it, or rather to detect the part of it useful to us, it will not be necessary to penetrate the mechanism; the law of great numbers will suffice. Let us take again the instance of the propagation of heat. Every molecule emits rays toward every neighboring molecule. According to what law, we do not need to know. If we should make any supposition in regard to this, it would be a neutral hypothesis and consequently useless and incapable of verification.

And, in fact, by the action of averages and thanks to the symmetry of the medium, all the differences are leveled down, and whatever hypothesis may be made, the result is always the same.

The same circumstance is presented in the theory of electricity and in that of capillarity. The neighboring molecules attract and repel one another. We do not need to know according to what law; it is enough for us that this attraction is sensible only at small distances, that the molecules are very numerous, that the medium is symmetrical, and we shall only have to let the law of great numbers act.

Here again the simplicity of the elementary phenomenon was hidden under the complexity of the resultant observable phenomenon; but, in its turn, this simplicity was only apparent, and concealed a very complex mechanism. The best means of arriving at the elementary phenomenon would evidently be experiment. We ought by experimental contrivance to dissociate the complex sheaf that nature offers to our researches, and to study with care the elements as much isolated as possible. For example, natural white light would be decomposed into monochromatic lights by the aid of the prism, and into polarized light by the aid of the polarizer.

Unfortunately that is neither always possible nor always sufficient, and sometimes the mind must outstrip experiment. I shall cite only one example, which has always struck me forcibly. If I decompose white light, I shall be able to isolate a small part of the spectrum, but however small it may be, it will retain a certain breadth. Likewise the natural lights, called *monochromatic*, give us a very narrow line, but not, however, infinitely narrow. It might be supposed that by studying experimentally the properties of these natural lights, by working with finer and finer lines of the spectrum, and by passing at last to the limit, so to speak, we should succeed in learning the properties of a light *strictly* monochromatic.

That would not be accurate. Suppose that two rays emanate from the same source, that we polarize them first in two perpendicular planes, then bring them back to the same plane of polarization, and try to make them interfere. If the light were strictly monochromatic, they would interfere. With our lights, which are nearly monochromatic, there will be no interference, and that no matter how narrow the line. In order to be otherwise it would have to be several million times as narrow as the finest known lines.

Here, then, the passage to the limit would have deceived us. The mind must outstrip the experiment, and if it has done so with success, it is because it has allowed itself to be guided by the instinct of simplicity. The knowledge of the elementary fact enables us to put the problem in an equation. Nothing remains but to deduce from this by combination the complex fact that can be observed and verified. This is what is called *integration*, and is the business of the mathematician.

It may be asked why, in physical sciences, generalization so readily takes the mathematical form. The reason is now easy to see. It is not only because we have numerical laws to express; it is because the observable phenomenon is due to the superposition of a great number of elementary phenomena *all alike*. Thus quite naturally are introduced differential equations.

It is not enough that each elementary phenomenon obeys simple laws; all those to be combined must obey the same law. Then only can the intervention of mathematics be of use; mathematics teaches us in fact to combine like with like. Its aim is to learn the result of a combination without needing to go over the combination piece by piece. If we have to repeat several times the same operation, it enables us to avoid this repetition by telling us in advance the result of it by a sort of induction. I have explained this above, in the chapter on mathematical reasoning.

But, for this, all the operations must be alike. In the opposite case, it would evidently be necessary to resign ourselves to doing them in reality one after another, and mathematics would become useless. It is then thanks to the approximate homogeneity of the matter studied by physicists, that mathematical physics could be born.

In the natural sciences, we no longer find these conditions: homogeneity, relative independence of remote parts, simplicity of the elementary fact; and this is why naturalists are obliged to resort to other methods of generalization.

13

Mathematical Creation
Henri Poincaré

The genesis of mathematical creation is a problem which should intensely interest the psychologist. It is the activity in which the human mind seems to take least from the outside world, in which it acts or seems to act only of itself and on itself, so that in studying the procedure of geometric thought we may hope to reach what is most essential in man's mind.

This has long been appreciated, and some time back the journal called *L'enseignement mathématique*, edited by Laisant and Fehr, began an investigation of the mental habits and methods of work of different mathematicians. I had finished the main outlines of this article when the results of that inquiry were published, so I have hardly been able to utilize them and shall confine myself to saying that the majority of witnesses confirm my conclusions; I do not say all, for when the appeal is to universal suffrage unanimity is not to be hoped.

A first fact should surprise us, or rather would surprise us if we were not so used to it. How does it happen there are people who do not understand mathematics? If mathematics invokes only the rules of logic, such as are accepted by all normal minds; if its evidence is based on principles common to all men, and that none could deny without being mad, how does it come about that so many persons are here refractory?

That not every one can invent is nowise mysterious. That not every one can retain a demonstration once learned may also pass. But that not every one can understand mathematical reasoning when explained appears very surprising when we think of it. And

* Reprinted from Henri Poincaré,"Hypotheses in Physics" in *The Foundations of Science* (New York: The Science Press, 1921) 383-394. Translated by George Bruce Halstead; public domain.

yet those who can follow this reasoning only with difficulty are in the majority: that is undeniable, and will surely not be gainsaid by the experience of secondary-school teachers.

And further: how is error possible in mathematics? A sane mind should not be guilty of a logical fallacy, and yet there are very fine minds who do not trip in brief reasoning such as occurs in the ordinary doings of life, and who are incapable of following or repeating without error the mathematical demonstrations which are longer, but which after all are only an accumulation of brief reasonings wholly analogous to those they make so easily. Need we add that mathematicians themselves are not infallible?

The answer seems to me evident. Imagine a long series of syllogisms, and that the conclusions of the first serve as premises of the following: we shall be able to catch each of these syllogisms, and it is not in passing from premises to conclusion that we are in danger of deceiving ourselves. But between the moment in which we first meet a proposition as conclusion of one syllogism, and that in which we reencounter it as premise of another syllogism occasionally some time will elapse, several links of the chain will have unrolled; so it may happen that we have forgotten it, or worse, that we have forgotten its meaning. So it may happen that we replace it by a slightly different proposition, or that, while retaining the same enunciation, we attribute to it a slightly different meaning, and thus it is that we are exposed to error.

Often the mathematician uses a rule. Naturally he begins by demonstrating this rule; and at the time when this proof is fresh in his memory he understands perfectly its meaning and its bearing, and he is in no danger of changing it. But subsequently he trusts his memory and afterward only applies it in a mechanical way; and then if his memory fails him, he may apply it all wrong. Thus it is, to take a simple example, that we sometimes make slips in calculation because we have forgotten our multiplication table.

According to this, the special aptitude for mathematics would be due only to a very sure memory or to a prodigious force of attention. It would be a power like that of the whist-player who remembers the cards played; or, to go up a step, like that of the chess-player who can visualize a great number of combinations and hold them in his memory. Every good mathematician ought to be a good chess-player, and inversely; likewise he should be a good computer. Of course that sometimes happens; thus Gauss was at the same time a geometer of genius and a very precocious and accurate computer.

But these are exceptions; or rather I err; I can not call them exceptions without the exceptions being more than the rule. Gauss it is, on the contrary, who was an exception. As for myself, I must confess, I am absolutely incapable even of adding without mistakes. In the same way I should be but a poor chessplayer; I would perceive that by a certain play I should expose myself to a certain danger; I would pass in review several other plays, rejecting them for other reasons, and then finally I should make the move first examined, having meantime forgotten the danger I had foreseen.

In a word, my memory is not bad, but it would be insufficient to make me a good chess-player. Why then does it not fail me in a difficult piece of mathematical reasoning where most chessplayers would lose themselves? Evidently because it is guided by the general march of the reasoning. A mathematical demonstration is not a simple juxtaposition of syllogisms, it is syllogisms *placed in a certain order*, and the order in which these elements are placed is much more important than the elements themselves. If I have the feeling, the intuition, so to speak, of this order, so as to perceive at a glance the reasoning as a whole, I need no longer fear lest I forget one of the elements, for each of them will take its allotted place in the array, and that without any effort of memory on my part.

It seems to me then, in repeating a reasoning learned, that I could have invented it. This is often only an illusion; but even then, even if I am not so gifted as to create it by myself, I myself re-invent it in so far as I repeat it. We know that this feeling, this intuition of mathematical order, that makes us divine hidden harmonies and relations, can not be possessed by every one. Some will not have either this delicate feeling so difficult to define, or a strength of memory and attention beyond the ordinary, and then they will be absolutely incapable of understanding higher mathematics. Such are the majority. Others will have this feeling only in a slight degree, but they will be gifted with an uncommon memory and a great power of attention. They will learn by heart the details one after another; they can understand mathematics and sometimes make applications, but they cannot create. Others, finally, will possess in a less or greater degree the special intuition referred to, and then not only can they understand mathematics even if their memory is nothing extraordinary, but they may become creators and try to invent with more or less success according as this intuition is more or less developed in them.

In fact, what is mathematical creation? It does not consist in making new combinations with mathematical entities already known. Any one could do that, but the combinations so made would be infinite in number and most of them absolutely without interest. To create consists precisely in not making useless combinations and in making those which are useful and which are only a small minority. Invention is discernment, choice.

How to make this choice I have before explained; the mathematical facts worthy of being studied are those which, by their analogy with other facts, are capable of leading us to the knowledge of a mathematical law just as experimental facts lead us to the knowledge of a physical law. They are those which reveal to us unsuspected kinship between other facts, long known, but wrongly believed to be strangers to one another.

Among chosen combinations the most fertile will often be those formed of elements drawn from domains which are far apart. Not that I mean as sufficing for invention the bringing together of objects as disparate as possible; most combinations so formed would be entirely sterile. But certain among them, very rare, are the most fruitful of all.

To invent, I have said, is to choose; but the word is perhaps not wholly exact. It makes one think of a purchaser before whom are displayed a large number of samples, and who examines them, one after the other, to make a choice. Here the samples would be so numerous that a whole lifetime would not suffice to examine them. This is not the actual state of things. The sterile combinations do not even present themselves to the mind of the inventor. Never in the field of his consciousness do combinations appear that are not really useful, except some that he rejects but which have to some extent the characteristics of useful combinations. All goes on as if the inventor were an examiner for the second degree who would only have to question the candidates who had passed a previous examination.

But what I have hitherto said is what may be observed or inferred in reading the writings of the geometers, reading reflectively. It is time to penetrate deeper and to see what goes on in the very soul of the mathematician. For this, I believe, I can do best by recalling memories of my own. But I shall limit myself to telling how I wrote my first memoir on Fuchsian functions. I beg the reader's pardon; I am about to use some technical expressions, but they need not frighten him, for he is not obliged to understand them. I shall say, for example, that I have found the demonstration of such a theorem under such circumstances. This theorem will have a barbarous name, unfamiliar to many, but that is unimportant; what is of interest for the psychologist is not the theorem but the circumstances.

For fifteen days I strove to prove that there could not be any functions like those I have since called Fuchsian functions. I was then very ignorant; every day I seated myself at my work table, stayed an hour or two, tried a great number of combinations and reached no results. One evening, contrary to my custom, I drank black coffee and could not sleep. Ideas rose in crowds; I felt them collide until pairs interlocked, so to speak, making a stable combination. By the next morning I had established the existence of a class of Fuchsian functions, those which come from the hypergeometric series; I had only to write out the results, which took but a few hours.

Then I wanted to represent these functions by the quotient of two series; this idea was perfectly conscious and deliberate, the analogy with elliptic functions guided me. I asked myself what properties these series must have if they existed, and I succeeded without difficulty in forming the series I have called theta Fuchsian.

Just at this time I left Caen, where I was then living, to go on a geologic excursion under the auspices of the school of mines. The changes of travel made me forget my mathematical work. Having reached Coutances, we entered an omnibus to go some place or other. At the moment when I put my foot on the step the idea came to me, without anything in my former thoughts seeming to have paved the way for it, that the transformations I had used to define the Fuchsian functions were identical with those of non-Euclidean geometry. I did not verify the idea; I should not have had time, as, upon taking my seat in the omnibus, I went on with a conversation already commenced, but I felt a perfect certainty. On my return to Caen, for conscience's sake I verified the result at my leisure.

Then I turned my attention to the study of some arithmetical questions apparently without much success and without a suspicion of any connection with my preceding researches. Disgusted with my failure, I went to spend a few days at the seaside, and thought of something else. One morning, walking on the bluff, the idea came to me, with just the same characteristics of brevity, suddenness and immediate certainty, that the arithmetic transformations of indeterminate ternary quadratic forms were identical with those of non-Euclidean geometry.

Returned to Caen, I meditated on this result and deduced the consequences. The example of quadratic forms showed me that there were Fuchsian groups other than those corresponding to the hypergeometric series; I saw that I could apply to them the theory of theta-Fuchsian series and that consequently there existed Fuchsian functions other than those from the hypergeometric series, the ones I then knew. Naturally I set myself to form all these functions. I made a systematic attack upon them and carried all the outworks, one after another. There was one however that still held out, whose fall would involve that of the whole place. But all my efforts only served at first the better to show me the difficulty, which indeed was something. All this work was perfectly conscious.

Thereupon I left for Mont-Valerien, where I was to go through my military service; so I was very differently occupied. One day, going along the street, the solution of the difficulty which had stopped me suddenly appeared to me. I did not try to go deep into it immediately, and only after my service did I again take up the question. I had all the elements and had only to arrange them and put them together. So I wrote out my final memoir at a single stroke and without difficulty.

I shall limit myself to this single example; it is useless to multiply them. In regard to my other researches I would have to say analogous things, and the observations of other mathematicians given in *L'enseignement mathematique* would only confirm them.

Most striking at first is this appearance of sudden illumination, a manifest sign of long, unconscious prior work. The role of this unconscious work in mathematical invention appears to me incontestable, and traces of it would be found in other cases where it is less evident. Often when one works at a hard question, nothing good is accomplished at the first attack. Then one takes a rest, longer or shorter, and sits down anew to the work. During the first half-hour, as before, nothing is found, and then all of a sudden the decisive idea presents itself to the mind. It might be said that the conscious work has been more fruitful because it has been interrupted and the rest has given back to the mind its force and freshness. But it is more probable that this rest has been filled out with unconscious work and that the result of this work has afterward revealed itself to the geometer just as in the cases I have cited; only the revelation, instead of coming during a walk or a journey, has happened during a period of conscious work, but independently of this work which plays at most a role of excitant, as if it were the goad stimulating the results already reached during rest, but remaining unconscious, to assume the conscious form.

There is another remark to be made about the conditions of this unconscious work: it is possible, and of a certainty it is only fruitful, if it is on the one hand preceded and on the other hand followed by a period of conscious work. These sudden inspirations (and the examples already cited sufficiently prove this) never happen except after some days of voluntary effort which has appeared absolutely fruitless and whence nothing good seems to have come, where the way taken seems totally astray. These efforts then have not been as sterile as one thinks; they have set agoing the unconscious machine and without them it would not have moved and would have produced nothing.

The need for the second period of conscious work, after the inspiration, is still easier to understand. It is necessary to put in shape the results of this inspiration, to deduce from them the immediate consequences, to arrange them, to word the demonstrations, but above all is verification necessary. I have spoken of the feeling of absolute certitude accompanying the inspiration; in the cases cited this feeling was no deceiver, nor is it usually. But do not think this a rule without exception; often this feeling deceives us without being any the less vivid, and we only find it out when we seek to put on foot the demonstration. I have especially noticed this fact in regard to ideas coming to me in the morning or evening in bed while in a semi-hypnagogic state.

Such are the realities; now for the thoughts they force upon us. The unconscious, or, as we say, the subliminal self plays an important role in mathematical creation; this follows from what we have said. But usually the subliminal self is considered as purely automatic. Now we have seen that mathematical work is not simply mechanical, that it could not be done by a machine, however perfect. It is not merely a question of applying rules, of making the most combinations possible according to certain fixed laws. The combinations so obtained would be exceedingly numerous, useless and cumbersome. The true work of the inventor consists in choosing among these combinations so as to eliminate the useless ones or rather to avoid the trouble of making them, and the rules which must guide this choice are extremely fine and delicate. It is almost impossible to state them precisely; they are felt rather than formulated. Under these conditions, how imagine a sieve capable of applying them mechanically?

A first hypothesis now presents itself: the subliminal self is in no way inferior to the conscious self; it is not purely automatic; it is capable of discernment; it has tact, delicacy; it knows how to choose, to divine. What do I say? It knows better how to divine than the conscious self, since it succeeds where that has failed. In a word, is not the subliminal self superior to the conscious self? You recognize the full importance of this question. Boutroux in a recent lecture has shown how it came up on a very different occasion, and what consequences would follow an affirmative answer. (See also, by the same author, *Science et Religion*)

Is this affirmative answer forced upon us by the facts I have just given? I confess that, for my part, I should hate to accept it. Reexamine the facts then and see if they are not compatible with another explanation. It is certain that the combinations which present themselves to the mind in a sort of sudden illumination, after an unconscious working somewhat prolonged, are generally useful and fertile combinations, which seem the result of a first impression. Does it follow that the subliminal self, having divined by a delicate intuition that these combinations would be useful, has formed only these, or has it rather formed many others which were lacking in interest and have remained unconscious?

In this second way of looking at it, all the combinations would be formed in consequence of the automatism of the subliminal self, but only the interesting ones would break into the domain of consciousness. And this is still very mysterious. What is the cause that, among the thousand products of our unconscious activity, some are called to pass the threshold, while others remain below? Is it a simple chance which confers this privilege? Evidently not; among all the stimuli of our senses, for example, only the most intense fix our attention, unless it has been drawn to them by other causes. More generally the privileged unconscious phenomena, those susceptible of becoming conscious, are those which, directly or indirectly, affect most profoundly our emotional sensibility.

It may be surprising to see emotional sensibility invoked *à propos* of mathematical demonstrations which, it would seem, can interest only the intellect. This would be to forget the feeling of mathematical beauty, of the harmony of numbers and forms, of geometric elegance. This is a true esthetic feeling that all real mathematicians know, and surely it belongs to emotional sensibility.

Now, what are the mathematic entities to which we attribute this character of beauty and elegance, and which are capable of developing in us a sort of esthetic emotion? They are those whose elements are harmoniously disposed so that the mind without effort can embrace their totality while realizing the details. This harmony is at once a satisfaction of our esthetic needs and an aid to the mind, sustaining and guiding. And at the same time, in putting under our eyes a well-ordered whole, it makes us foresee a mathematical law. Now, as we have said above, the only mathematical facts worthy of fixing our attention and capable of being useful are those which can teach us a mathematical law. So that we reach the following conclusion: The useful combinations are precisely the most beautiful, I mean those best able to charm this special sensibility that all mathematicians know, but of which the profane are so ignorant as often to be tempted to smile at it.

What happens then? Among the great numbers of combinations blindly formed by the subliminal self, almost all are without interest and without utility; but just for that reason they are also without effect upon the esthetic sensibility. Consciousness will never know them; only certain ones are harmonious, and, consequently, at once useful and beautiful. They will be capable of touching this special sensibility of the geometer of which I have just spoken, and which, once aroused, will call our attention to them, and thus give them occasion to become conscious.

This is only a hypothesis, and yet here is an observation which may confirm it: when a sudden illumination seizes upon the mind of the mathematician, it usually happens that it does not deceive him, but it also sometimes happens, as I have said, that it does not stand the test of verification; well, we almost always notice that this false idea, had it been true, would have gratified our natural feeling for mathematical elegance.

Thus it is this special esthetic sensibility which plays the role of the delicate sieve of which I spoke, and that sufficiently explains why the one lacking it will never be a real creator.

Yet all the difficulties have not disappeared. The conscious self is narrowly limited, and as for the subliminal self we know not its limitations, and this is why we are not too reluctant in supposing that it has been able in a short time to make more different combinations than the whole life of a conscious being could encompass. Yet these limitations exist. Is it likely that it is able to form all the possible combinations, whose number would frighten the imagination? Nevertheless that would seem necessary, because if it produces only a small part of these combinations, and if it makes them at random, there would he small chance that *the good*, the one we should choose, would be found among them.

Perhaps we ought to seek the explanation in that preliminary period of conscious work which always precedes all fruitful unconscious labor. Permit me a rough comparison. Figure the future elements of our combinations as something like the hooked atoms of Epicurus. During the complete repose of the mind, these atoms are motionless, they are, so to speak, hooked to the wall; so this complete rest may be indefinitely prolonged without the atoms meeting, and consequently without any combination between them.

On the other hand, during a period of apparent rest and unconscious work, certain of them are detached from the wall and put in motion. They flash in every direction through the space (I was about to say the room) where they are enclosed, as would, for example, a swarm of gnats or, if you prefer a more learned comparison, like the molecules of gas in the kinematic theory of gases. Then their mutual impacts may produce new combinations.

What is the role of the preliminary conscious work? It is evidently to mobilize certain of these atoms, to unhook them from the wall and put them in swing. We think we have done no good, because we have moved these elements a thousand different ways in seeking to assemble them, and have found no satisfactory aggregate. But, after this shaking up imposed upon them by our will, these atoms do not return to their primitive rest. They freely continue their dance.

Now, our will did not choose them at random; it pursued a perfectly determined aim. The mobilized atoms are therefore not any atoms whatsoever; they are those from which we might reasonably expect the desired solution. Then the mobilized atoms undergo impacts which make them enter into combinations among themselves or with other atoms at rest which they struck against in their course. Again I beg pardon, my comparison is very rough, but I scarcely know how otherwise to make my thought understood.

However it may be, the only combinations that have a chance of forming are those where at least one of the elements is one of those atoms freely chosen by our will. Now, it is evidently among these that is found what I called the *good combination*. Perhaps this is a way of lessening the paradoxical in the original hypothesis.

Another observation. It never happens that the unconscious work gives us the result of a somewhat long calculation *all made*, where we have only to apply fixed rules. We might think the wholly automatic subliminal self particularly apt for this sort of work, which is in a way exclusively mechanical. It seems that thinking in the evening upon the factors of a multiplication we might hope to find the product ready made upon our awakening, or again that an algebraic calculation, for example a verification, would be made unconsciously. Nothing of the sort, as observation proves. All one may hope from these inspirations, fruits of unconscious work, is a point of departure for such calculations. As for the calculations themselves, they must be made in the second period of conscious work, that which follows the inspiration, that in which one verifies the results of this inspiration and deduces their consequences. The rules of these calculations are strict and complicated. They require discipline, attention, will, and therefore consciousness. In the subliminal self, on the contrary, reigns what I should call liberty, if we might give this name to the simple absence of discipline and to the disorder born of chance. Only, this disorder itself permits unexpected combinations.

I shall make a last remark: when above I made certain personal observations, I spoke of a night of excitement when I worked in spite of myself. Such cases are frequent, and it is not necessary that the abnormal cerebral activity be caused by a physical excitant as in that I mentioned. It seems, in such cases, that one is present at his own unconscious work, made partially perceptible to the over-excited consciousness, yet without having changed its nature. Then we vaguely comprehend what distinguishes the two mechanisms or, if you wish, the working methods of the two egos. And the psychologic observations I have been able thus to make seem to me to confirm in their general outlines the views I have given.

Surely they have need of it, for they are and remain in spite of all very hypothetical: the interest of the questions is so great that I do not repent of having submitted them to the reader.

14

The Part Played by Accident in Invention and Discovery
Ernst Mach

Ernst Mach (February 18, 1838 – February 19, 1916)—was an Austrian physicist best known for his discovery of shock waves created by objects moving through a medium with a velocity faster than the velocity of sound in that medium. It is from this discovery that the 'Mach Number', the ratio of an object's velocity to the speed of sound, was coined. Mach often wrote on philosophical topics and while prescient on many fronts, such as a lack of satisfaction with Newtonian mechanics, he was blind on others such as his insistence that atoms were not real. This article is his description on how chance plays a large part in many scientific discoveries.

•Inaugural lecture delivered on assuming the Professorship of the History and Theory of Inductive Science in the University of Vienna, October 21, 1895.

It is characteristic of the naive and sanguine beginnings of thought in youthful men and nations, that all problems are held to be soluble and fundamentally intelligible on the first appearance of success. The sage of Miletus, on seeing plants take their rise from moisture, believed he had comprehended the whole of nature, and he of Samos, on discovering that definite numbers corresponded to the lengths of harmonic strings, imagined he could exhaust the nature of the world by means of numbers. Philosophy and science in such periods are blended. Wider experience, however, speedily discloses the error of such a course, gives rise to criticism, and leads to the division and ramification of the sciences.

---

* Reprinted from Ernst Mach,"The Part Played by Accident in Invention and Discovery" in *Popular Scientific Lectures 3rd Edition* (Chicago: Open Court Publishing, 1898) 259-281; public domain

At the same time, the necessity of a broad and general view of the world remains; and to meet this need philosophy parts company with special inquiry. It is true, the two are often found united in gigantic personalities. But as a rule their ways diverge more and more widely from each other. And if the estrangement of philosophy from science can reach a point where data unworthy of the nursery are not deemed too scanty as foundations of the world, on the other hand the thorough-paced specialist may go to the extreme of rejecting point-blank the possibility of a broader view, or at least of deeming it superfluous, forgetful of Voltaire's apophthegm, nowhere more applicable than here, *Le superflu—chose très nécessaire* (the superfluous is very necessary).

It is true, the history of philosophy, owing to the insufficiency of its constructive data, is and must be largely a history of error. But it would be the height of ingratitude on our part to forget that the seeds of thoughts which still fructify the soil of special research, such as the theory of irrationals, the conceptions of conservation, the doctrine of evolution, the idea of specific energies, and so forth, may be traced back in distant ages to philosophical sources. Furthermore, to have deferred or abandoned the attempt at a broad philosophical view of the world from a full knowledge of the insufficiency of our materials, is quite a different thing from never having undertaken it at all. The revenge of its neglect, moreover, is constantly visited upon the specialist by his committal of the very errors which philosophy long ago exposed. As a fact, in physics and physiology, particularly during the first half of this century, are to be met intellectual productions which for naive simplicity are not a jot inferior to those of the Ionian school, or to the Platonic ideas, or to that much reviled ontological proof.

Latterly, there has been evidence of a gradual change in the situation. Recent philosophy has set itself more modest and more attainable ends; it is no longer inimical to special inquiry; in fact, it is zealously taking part in that inquiry. On the other hand, the special sciences, mathematics and physics, no less than philology, have become eminently philosophical. The material presented is no longer accepted uncritically. The glance of the inquirer is bent upon neighboring fields, whence that material has been derived. The different special departments are striving for closer union, and gradually the conviction is gaining ground that philosophy can consist only of mutual, complemental criticism, interpenetration, and union of the special sciences into a consolidated whole. As the blood in nourishing the body separates into countless capillaries, only to be collected again and to meet in the heart, so in the science of the future all the rills of knowledge will gather more and more into a common and undivided stream.

It is this view—not an unfamiliar one to the present generation—that I purpose to advocate. Entertain no hope, or rather fear, that I shall construct systems for you. I shall remain a natural inquirer. Nor expect that it is my intention to skirt all the fields of natural inquiry. I can attempt to be your guide only in that branch which is familiar to me, and even there I can assist in the furtherment of only a small portion of the allotted task. If I shall succeed in rendering plain to you the relations of physics, psychology, and the theory of knowledge, so that you may draw from each profit and light, redounding to the advantage of each, I shall regard my work as not having been in vain. Therefore, to illustrate by an example how, consonantly with my powers and views, I conceive such inquiries should be conducted, I shall treat to-day, in the form of a brief sketch, of the following special and limited subject—of *the part which accidental circumstances play in the development of inventions and discoveries.*

When we Germans say of a man that he was not the inventor of gunpowder[1], we impliedly cast a grave suspicion on his abilities. But the expression is not a felicitous one, as there is probably no invention in which deliberate thought had a smaller, and pure luck a larger, share than in this. It is well to ask, are we justified in placing a low estimate on the achievement of an inventor because accident has assisted him in his work? Huygens, whose discoveries and inventions are justly sufficient to entitle him to an opinion in such matters, lays great emphasis on this factor. He asserts that a man capable of inventing the telescope without the concurrence of accident must have been gifted with superhuman genius[2].

---

[1] The phrase is "Er hat das pulver nicht erfunden"

[2] But if anyone had ever existed of such great diligence, that he proved able to derive this matter from the principles of nature and geometry, then, I believe, it should be said that his intellectual powers were

A man living in the midst of civilisation finds himself surrounded by a host of marvellous inventions, considering none other than the means of satisfying the needs of daily life. Picture such a man transported to the epoch preceding the invention of these ingenious appliances, and imagine him undertaking in a serious manner to comprehend their origin. At first the intellectual power of the men capable of producing such marvels will strike him as incredible, or, if we adopt the ancient view, as divine. But his astonishment is considerably allayed by the disenchanting yet elucidative revelations of the history of primitive culture, which to a large extent prove that these inventions took their rise very slowly and by imperceptible degrees.

A small hole in the ground with fire kindled in it constituted the primitive stove. The flesh of the quarry, wrapped with water in its skin, was boiled by contact with heated stones. Cooking by stones was also done in wooden vessels. Hollow gourds were protected from the fire by coats of clay. Thus, from the burned clay accidentally originated the enveloping pot, which rendered the gourd superfluous, although for a long time thereafter the clay was still spread over the gourd, or pressed into woven wicker-work before the potter's art assumed its final independence. Even then the wicker-work ornament was retained, as a sort of attest of its origin.

We see, thus, it is by accidental circumstances, or by such as lie without our purpose, foresight, and power, that man is gradually led to the acquaintance of improved means of satisfying his wants. Let the reader picture to himself the genius of a man who could have foreseen without the help of accident that clay handled in the ordinary manner would produce a useful cooking utensil! The majority of the inventions made in the early stages of civilisation, including language, writing, money, and the rest, could not have been the product of deliberate methodical reflexion for the simple reason that no idea of their value and significance could have been had except from practical use. The invention of the bridge may have been suggested by the trunk of a tree which had

---

above those of ordinary mortals. But we are so far removed from this situation, that even the most erudite scholars have not as yet been able to give an adequate explanation of the way this device works, which was discovered by chance. Christian Huygens, On Telescopes.

fallen athwart a mountain-torrent; that of the tool by the use of a stone accidentally taken into the hand to crack nuts. The use of fire probably started in and was disseminated from regions where volcanic eruptions, hot springs, and burning jets of natural gas afforded opportunity for quietly observing and turning to practical account the properties of fire. Only after that had been done could the significance of the fire-drill be appreciated, an instrument which was probably discovered from boring a hole through a piece of wood. The suggestion of a distinguished inquirer that the invention of the fire-drill originated on the occasion of a religious ceremony is both fantastic and incredible. And as to the use of fire, we should no more attempt to derive that from the invention of the fire-drill than we should from the invention of sulphur matches. Unquestionably the opposite course was the real one[3].

Similar phenomena, though still largely veiled in obscurity, mark the initial transition of nations from a hunting to a nomadic life and to agriculture[4]. We shall not multiply examples, but content ourselves with the remark that the same phenomena recur in historical times, in the ages of great technical inventions, and, further, that regarding them the most whimsical notions have been circulated—notions which ascribe to accident an unduly exaggerated part, and one which in a psychological respect is absolutely impossible. The observation of steam escaping from a tea-kettle and of the clattering of the lid is supposed to have led to the invention of the steam-engine. Just think of the gap between this spectacle and the conception of the performance of great mechanical work by steam, for a man totally ignorant of the steam engine! Let us suppose, however, that an engineer, versed in the practical construction of pumps, should accidentally dip into water an inverted bottle that had been filled with steam for drying and still retained its steam. He would see the water rush violently into the bottle, and the idea would very naturally suggest itself of founding on this experience a convenient and useful atmospheric steam-pump, which by imperceptible degrees, both psychologically pos-

[3] I must not be understood as saying that the fire-drill has played no part in the worship of fire or of the sun.

[4] Compare on this point the extremely interesting remarks of Dr. Paul Carus in his Philosophy of the Tool, Chicago, 1893.

sible and immediate, would then undergo a natural and gradual transformation into Watt's steam-engine.

But granting that the most important inventions are brought to man's notice accidentally and in ways that are beyond his foresight, yet it does not follow that accident alone is sufficient to produce an invention. The part which man plays is by no means a passive one. Even the first potter in the primeval forest must have felt some stirrings of genius within him. In all such cases, the inventor is obliged *to take note* of the new fact, he must discover and grasp its advantageous feature, and must have the power to turn that feature to account in the realisation of his purpose. He must *isolate* the new feature, impress it upon his memory, unite and interweave it with the rest of his thought; in short, he must possess the capacity to *profit by experience.*

The capacity to profit by experience might well be set up as a test of intelligence. This power varies considerably in men of the same race, and increases enormously as we advance from the lower animals to man. The former are limited in this regard almost entirely to the reflex actions which they have inherited with their organism, they are almost totally incapable of individual experience, and considering their simple wants are scarcely in need of it. The ivory-snail (*Eburna spirata*) never learns to avoid the carnivorous Actinia, no matter how often it may wince under the latter's shower of needles, apparently having no memory for pain whatever[5]. A spider can be lured forth repeatedly from its hole by touching its web with a tuning-fork. The moth plunges again and again into the flame which has burnt it. The humming-bird hawk-moth[6] dashes repeatedly against the painted roses of the wall-paper, like the unhappy and desperate thinker who never wearies of attacking the same insoluble chimerical problem. As aimlessly almost as Maxwell's gaseous molecules and in the same unreasoning manner common flies in their search for light and air stream against the glass pane of a half-opened window and remain there from sheer inability to find their way around the narrow frame. But a pike

[5] Möbius, Naturwissenschaftlicher Verein für Schleswig-Holstein, Kiel, 1893, p. 113 et seq.
[6] I am indebted for this observation to Professor Hatscheck.

separated from the minnows of his aquarium by a glass partition, learns after the lapse of a few months, though only after having butted himself half to death, that he cannot attack these fishes with impunity. What is more, he leaves them in peace even after the removal of the partition, though he will bolt a strange fish at once. Considerable memory must be attributed to birds of passage, a memory which, probably owing to the absence of disturbing thoughts, acts with the precision of that of some idiots. Finally, the susceptibility to training evinced by the higher vertebrates is indisputable proof of the ability of these animals to profit by experience.

A powerfully developed *mechanical* memory, which recalls vividly and faithfully old situations, is sufficient for avoiding definite particular dangers, or for taking advantage of definite particular opportunities. But more is required for the development of *inventions.* More extensive chains of images are necessary here, the excitation by mutual contact of widely different trains of ideas, a more powerful, more manifold, and richer connexion of the contents of memory, a more powerful and impressionable psychical life, heightened by use. A man stands on the bank of a mountain-torrent, which is a serious obstacle to him. He remembers that he has crossed just such a torrent before on the trunk of a fallen tree. Hard by trees are growing. He has often moved the trunks of fallen trees. He has also felled trees before, and then moved them. To fell trees he has used sharp stones. He goes in search of such a stone, and as the old situations that crowd into his memory and are held there in living reality by the definite powerful interest which he has in crossing just this torrent,—as these impressions are made to pass before his mind in the *inverse order* in which they were here evoked, he invents the bridge.

There can be no doubt but the higher vertebrates adapt their actions in some moderate degree to circumstances. The fact that they give no appreciable evidence of advance by the accumulation of inventions, is satisfactorily explained by a difference of degree or intensity of intelligence as compared with man; the assumption of a difference of kind is not necessary. A person who saves a little every day, be it ever so little, has an incalculable advantage over him who daily squanders that amount, or is unable to keep what he has accumulated. A slight quantitative difference in such things explains enormous differences of advancement.

The rules which hold good in prehistoric times also hold good in historical times, and the remarks made on invention may be applied almost without modification to discovery; for the two are distinguished solely by the use to which the new knowledge is put. In both cases the investigator is concerned with some *newly observed* relation of new or old properties, abstract or concrete. It is observed, for example, that a substance which gives a chemical reaction *A* is also the cause of a chemical reaction *B*. If this observation fulfils no purpose but that of furthering the scientist's insight, or of removing a source of intellectual discomfort, we have a discovery; but an invention, if in using the substance giving the reaction *A* to produce the desired reaction *B*, we have a practical end in view, and seek to remove a source of material discomfort. The phrase, *disclosure of the connexion of reactions*, is broad enough to cover discoveries and inventions in all departments. It embraces the Pythagorean proposition, which is a combination of a geometrical and an arithmetical reaction, Newton's discovery of the connexion of Kepler's motions with the law of the inverse squares, as perfectly as it does the detection of some minute but appropriate alteration in the construction of a tool, or of some appropriate change in the methods of a dyeing establishment.

The disclosure of new provinces of facts before unknown can only be brought about by accidental circumstances, under which are *remarked* facts that commonly go unnoticed. The achievement of the discoverer here consists in his *sharpened attention*, which detects the uncommon features of an occurrence and their determining conditions from their most evanescent marks[7], and discovers means of submitting them to exact and full observation. Under this head belong the first disclosures of electrical and magnetic phenomena, Grimaldi's observation of interference, Arago's discovery of the increased check suffered by a magnetic needle vibrating in a copper envelope as compared with that observed in a bandbox, Foucault's observation of the stability of the plane of vibration of a rod accidentally struck while rotating in a turning-

[7] Cf. Hoppe, Entdecken und Finden. 1870.

lathe, Mayer's observation of the increased redness of venous blood in the tropics, Kirchhoff's observation of the augmentation of the D-Line in the solar spectrum by the interposition of a sodium lamp, Schönbein's discovery of ozone from the phosphoric smell emitted on the disruption of air by electric sparks, and a host of others. All these facts, of which unquestionably many were *seen* numbers of times before they were *noticed*, are examples of the inauguration of momentous discoveries by accidental circumstances, and place the importance of strained attention in a brilliant light.

But not only is a significant part played in the beginning of an inquiry by co-operative circumstances beyond the foresight of the investigator; their influence is also active in its prosecution. Dufay, thus, whilst following up the behavior of *one* electrical state which he had assumed, discovers the existence of *two*. Fresnel learns by accident that the interference-bands received on ground glass are seen to better advantage in the open air. The diffraction-phenomenon of two slits proved to be considerably different from what Fraunhofer had anticipated, and in following up this circumstance he was led to the important discovery of grating-spectra. Faraday's induction-phenomenon departed widely from the initial conception which occasioned his experiments, and it is precisely this deviation that constitutes his real discovery.

Every man has pondered on some subject. Every one of us can multiply the examples cited, by less illustrious ones from his own experience. I shall cite but one. On rounding a railway curve once, I accidentally remarked a striking apparent inclination of the houses and trees. I inferred that the direction of the total resultant *physical* acceleration of the body reacts *physiologically* as the vertical. Afterwards, in attempting to inquire more carefully into this phenomenon, and this only, in a large whirling machine, the collateral phenomena conducted me to the sensation of angular acceleration, vertigo, Flouren's experiments on the section of the semicircular canals etc., from which gradually resulted views relating to sensations of direction which are also held by Breuer and Brown, which were at first contested on all hands, but are now regarded on many sides as correct, and which have been recently enriched by the interesting inquiries of Breuer concerning the *macula acustica*, and Kreidel's experiments with magnetically orientable Crustacea[8]. Not disregard of accident but a direct and purposeful employment of it advances research.

The more powerful the psychical connexion of the memory pictures is,—and it varies with the individual and the mood,—the more apt is the same accidental observation to be productive of results. Galileo knows that the air has weight; he also knows of the "resistance to a vacuum," expressed both in weight and in the height of a column of water. But the two ideas dwelt asunder in his mind. It remained for Torricelli to vary the specific gravity of the liquid measuring the pressure, and not till then was the air included in the list of pressure-exerting fluids. The reversal of the lines of the spectrum was seen repeatedly before Kirchhoff, and had been mechanically explained. But it was left for his penetrating vision to discern the evidence of the connexion of this phenomenon with questions of heat, and to him alone through persistent labor was revealed the sweeping significance of the fact for the mobile equilibrium of heat. Supposing, then, that such a rich organic connexion of the elements of memory exists, and is the prime distinguishing mark of the inquirer, next in importance certainly is that *intense interest* in a definite object, in a definite idea, which fashions advantageous combinations of thought from elements before disconnected, and obtrudes that idea into every observation made, and into every thought formed, making it enter into relationship with all things. Thus Bradley, deeply engrossed with the subject of aberration, is led to its solution by an exceedingly unobtrusive experience in crossing the Thames. It is permissible, therefore, to ask whether accident leads the discoverer, or the discoverer accident, to a successful outcome in scientific quests.

No man should dream of solving a great problem unless he is so thoroughly saturated with his subject that everything else sinks into comparative insignificance. During a hurried meeting with Mayer in Heidelberg once, Jolly remarked, with a rather dubious implication, that if Mayer's theory were correct water could be warmed by shaking. Mayer went away without a word of reply.

---

[8] See the lecture "Sensations of Orientation"

Several weeks later, and now unrecognised by Jolly[9], he rushed into the latter's presence exclaiming: "Es ischt aso!" (It is so, it is so!) It was only after considerable explanation that Jolly found out what Mayer wanted to say. The incident needs no comment.

A person deadened to sensory impressions and given up solely to the pursuit of his own thoughts, may also light on an idea that will divert his mental activity into totally new channels. In such cases it is a psychical accident, an intellectual experience, as distinguished from a physical accident, to which the person owes his discovery—a discovery which is here made "deductively" by means of mental copies of the world, instead of experimentally. *Purely* experimental inquiry, moreover, does not exist, for, as Gauss says, virtually we always experiment with our thoughts. And it is precisely that constant, corrective interchange or intimate union of experiment and deduction, as it was cultivated by Galileo in his *Dialogues* and by Newton in his *Optics*, that is the foundation of the benign fruitfulness of modern scientific inquiry as contrasted with that of antiquity, where observation and reflexion ofttimes pursued their respective courses like two strangers.

We have to wait for the appearance of a favorable physical accident. The movement of our thoughts obeys the law of association. In the case of meagre experience the result of this law is simply the mechanical reproduction of definite sensory experiences. On the other hand, if the psychical life is subjected to the incessant influences of a powerful and rich experience, then every representative element in the mind is connected with so many others that the actual and natural course of the thoughts is easily influenced and determined by insignificant circumstances, which accidentally are decisive. Hereupon, the process termed imagination produces its protean and infinitely diversified forms. Now what can we do to guide this process, seeing that the combinatory law of the images is without our reach? Rather let us ask, what influence can a powerful and constantly recurring idea exert on the movement of our thoughts? According to what has preceded, the

[9] This story was related to me by Jolly, and subsequently repeated in a letter from him.

answer is involved in the question itself. The *idea* dominates the thought of the inquirer, not the latter the former.

Let us see, now, if we can acquire a profounder insight into the process of discovery. The condition of the discoverer is, as James has aptly remarked, not unlike the situation of a person who is trying to remember something that he has forgotten. Both are sensible of a gap, and have only a remote presentiment of what is missing. Suppose I meet in a company a well-known and affable gentleman whose name I have forgotten, and who to my horror asks to be introduced to some one. I set to work according to Lichtenberg's rule, and run down the alphabet in search of the initial letter of his name. A vague sympathy holds me at the letter *G*. Tentatively I add the second letter and am arrested at *e*, and long before I have tried the third letter *r*, the name "Gerson" sounds sonorously upon my ear, and my anguish is gone. While taking a walk I meet a gentleman from whom I receive a communication. On returning home, and in attending to weightier affairs, the matter slips my mind. Moodily, but in vain, I ransack my memory. Finally I observe that I am going over my walk again in thought. On the street corner in question the selfsame gentleman stands before me and repeats his communication. In this process are successively recalled to consciousness all the percepts which were connected with the percept that was lost, and with them, finally, that, too, is brought to light. In the first case—where the experience had already been made and is permanently impressed on our thought—a *systematic* procedure is both possible and easy, for we know that a name must be composed of a limited number of sounds. But at the same time it should be observed that the labor involved in such a combinatorial task would be enormous if the name were long and the responsiveness of the mind weaker.

It is often said, and not wholly without justification, that the scientist has solved a *riddle*. Every problem in geometry may be clothed in the garb of a *riddle*. Thus: "What thing is that *M* which has the properties *A*, *B*, *C*?" "What circle is that which touches the straight lines *A*, *B*, but touches *B* in the point *C*?" The first two conditions marshal before the imagination the group of circles whose centres lie in the line of symmetry of *A*, *B*. The third condition reminds us of all the circles having centres in the straight line that stands at right angles to *B* in *C*. The *common* term, or common terms, of the two groups of images solves the riddle—satisfies the problem. Puzzles dealing with things or words induce similar processes, but the memory in such cases is exerted in many directions and more varied and less clearly ordered provinces of ideas are surveyed. The difference between the situation of a geometer who has a construction to make, and that of an engineer, or a scientist, confronted with a problem, is simply this, that the first moves in a field with which he is thoroughly acquainted, whereas the two latter are obliged to familiarise themselves with this field subsequently, and in a measure far transcending what is commonly required. In this process the mechanical engineer has at least always a definite goal before him and definite means to accomplish his aim, whilst in the case of the scientist that aim is in many instances presented only in vague and general outlines. Often the very formulation of the riddle devolves on him. Frequently it is not until the aim has been reached that the broader outlook requisite for systematic procedure is obtained. By far the larger portion of his success, therefore, is contingent on luck and instinct. It is immaterial, so far as its character is concerned, whether the process in question is brought rapidly to a conclusion in the brain of one man, or whether it is spun out for centuries in the minds of a long succession of thinkers. The same relation that a word solving a riddle bears to that riddle is borne by the modern conception of light to the facts discovered by Grimaldi, Römer, Huygens, Newton, Young, Malus, and Fresnel, and only by the help of this slowly developed conception is our mental vision enabled to embrace the broad domain of facts in question.

A welcome complement to the discoveries which the history of civilisation and comparative psychology have furnished, is to be found in the confessions of great scientists *and* artists. Scientists and artists, we might say, for Liebig boldly declared there was no essential difference between the two. Are we to regard Leonardo da Vinci as a scientist or as an artist? If the artist builds up his work from a few motives, the scientist discovers the motives which permeate reality. If scientists like Lagrange or Fourier are in a certain measure artists in the presentation of their results, on the other hand, artists like Shakespeare or Ruysdael are scientists in the insight which must have preceded their creations.

Newton, when questioned about his methods of work, could give no other answer but that he was wont to ponder again and again on a subject; and similar utterances are accredited to D'Alembert and Helmholtz. Scientists and artists both recommend persistent labor. After the repeated survey of a field has afforded opportunity for the interposition of advantageous accidents, has rendered all the traits that suit with the mood or the dominant thought more vivid, and has gradually relegated to the background all things that are inappropriate, making their future appearance impossible; then from the teeming, swelling host of fancies which a free and high-flown imagination calls forth, suddenly that particular form arises to the light which harmonises perfectly with the ruling idea, mood, or design. Then it is that that which has resulted slowly as the result of a gradual selection, appears as if it were the outcome of a deliberate act of creation. Thus are to be explained the statements of Newton, Mozart, Richard Wagner, and others, when they say that thoughts, melodies, and harmonies had poured in upon them, and that they had simply retained the right ones. Undoubtedly, the man of genius, too, consciously or instinctively, pursues systematic methods wherever it is possible; but in his delicate presentiment he will omit many a task or abandon it after a hasty trial on which a less endowed man would squander his energies in vain. Thus, the genius accomplishes[10] in a brief space of time undertakings for which the life of an ordinary man would far from suffice. We shall hardly go astray if we regard genius as only a slight deviation from the average mental endowment—as possessing simply a greater sensitiveness of cerebral reaction and a greater swiftness of reaction. The men who, obeying their inner impulses, make sacrifices for an idea instead of advancing their material welfare, may appear to the full-blooded Philistine as fools; yet we shall scarcely adopt Lombroso's view,

[10] I do not know whether Swift's academy of schemers in Lagado, in which great discoveries and inventions were made by a sort of verbal game of dice was intended as a satire on Francis Bacon's method of making discoveries by means of huge synoptic tables constructed by scribes. It certainly would not have been ill-placed.

that genius is to be regarded as a disease, although it is unfortunately true that the sensitive brains and fragile constitutions succumb most readily to sickness.

The remark of C.G.J. Jacobi that mathematics is slow of growth and only reaches the truth by long and devious paths, that the way to its discovery must be prepared for long beforehand, and that then the truth will make its long-deferred appearance as if impelled by some divine necessity[11]—all this holds true of every science. We are astounded often to note that it required the combined labors of many eminent thinkers for a full century to reach a truth which it takes us only a few hours to master and which once acquired seems extremely easy to reach under the right sort of circumstances. To our humiliation we learn that even the greatest men are born more for life than for science. The extent to which even they are indebted to accident—to that singular conflux of the physical and the psychical life in which the continuous but yet imperfect and never-ending adaptation of the latter to the former finds its distinct expression—that has been the subject of our remarks to-day. Jacobi's poetical thought of a divine necessity acting in science will lose none of its loftiness for us if we discover in this necessity the same power that destroys the unfit and fosters the fit. For loftier, nobler, and more romantic than poetry is the truth and the reality.

[11]Branches of learning develop slowly and tardily. Late in the day after passing through various errors one arrives at the truth. All things must be made ready by means of lengthy and unremitting toil, in order to gain access to a new truth. Only then will it finally emerge at some definite point of time, driven by a sort of divine compulsion.
Quoted by Simony, In ein ringförmiges Band einen Knoten zu machen, Vienna, 1881, p. 41

15

The Art of Scientific Investigation: Chapter 3: Chance William Ian Beardmore Beveridge

"Chance favours only those who know how to court her."
—Charles Nicolle

Illustrations

It will be simpler to discuss the role of chance in research if we first consider some illustrative examples of discoveries in which it played a part. These anecdotes have been taken from sources believed to be authentic, and one reference is quoted for each although in many instances several sources have been consulted. Only ten are included in this section but seventeen others illustrating the role of chance are to be found in the Appendix. Pasteur's researches on fowl cholera were interrupted by the vacation, and when he resumed he encountered an unexpected obstacle. Nearly all the cultures had become sterile. He attempted to revive them by sub-inoculation into broth and injection into fowls. Most of the sub-cultures failed to grow and the birds were not affected, so he was about to discard everything and start afresh when he had the inspiration of re-inoculating the same fowls with a fresh culture. His colleague Duclaux relates:

> *To the surprise of all, and perhaps even of Pasteur, who was not expecting such success, nearly all these fowls withstood the inoculadon, although fresh fowls succumbed after the usual incubation period.*

This resulted in the recognition of the principle of immunisation with attenuated pathogens [1]. The most important method used in staining bacteria is that discovered by the Danish physician

---

* Reprinted from W.I.B. Beveridge, The Art of Scientific Investigation (New York: W.W. Norton, 1957) 27-40; 160-168 under

H.C. Gram. He described how he discovered the method fortuitously when trying to develop a double stain for kidney sections. Hoping to stain the nuclei violet and the tubules brown, he used gentian violet followed by iodine solution. Gram found that after this treatment the tissue was rapidly decolourised by alcohol but that certain bacteria remained blue-black. The gentian violet and iodine had unexpectedly reacted with each other and with a substance present in some bacteria and not others, thus providing not only a good stain but also a simple test which has proved of the greatest value in distinguishing different bacteria [2].

While engaged in studying the function of the pancreas in digestion in 1889 at Strasbourg, Professors von Mering and Minkowski removed that organ from a dog by operation. Later a laboratory assistant noticed that swarms of flies were attracted by the urine of the operated dog. He brought this to the attention of Minkowski, who analysed the urine and found sugar in it. It was this finding that led to our understanding of diabetes and its subsequent control by insulin [3]. More recently the Scotsman, Shaw Dunn, was investigating the cause of the kidney damage which follows a severe crush injury to a limb. Among other things he injected alloxan and he found that it caused necrosis of the islet tissue of the pancreas. This unexpected finding has provided a most useful tool in the study of diabetes [4].

The French physiologist, Charles Richet, was testing an extract of the tentacles of a sea anemone on laboratory animals to determine the toxic dose when he found that a small second dose given some time after the first was often promptly fatal. He was at first so astounded at this result that he could hardly believe that it was due to anything he had done. Indeed he said it was in spite of himself that he discovered induced sensitisation or anaphylaxis and that he would never have believed that it was possible." [3] Another manifestation of the same phenomenon was discovered independently by Sir Henry Dale. He was applying serum to strips of involuntary muscle taken from guinea-pigs when he encountered one that reacted violently to the application of horse serum. Seeking an explanation of this extraordinary observation he found that that guinea-pig had some time previously been injected with horse serum. [5]

It was the usual practice among physiologists to use physiological saline as a perfusion fluid during experiments on isolated frogs' hearts. By this means they could be kept beating for perhaps half an hour. Once at the London University College Hospital a physiologist was surprised and puzzled to find his frogs' hearts continued to beat for many hours. The only possible explanation he could think of was that it was a seasonal effect and this he actually suggested in a report. Then it was found that the explanation was that his laboratory assistant had used tap water instead of distilled water to make up the saline solution. With this clue it was easy to determine what salts in the tap water were responsible for the increased physiological activity. This was what led Sidney Ringer to develop the solution which bears his name and which has contributed so much to experimental physiology. [5]

Dr. H.E. Durham has left the following written account of the discovery of agglutination of bacteria by antiserum.

> *It was a memorable morning in November 1894, when we had all made ready with culture and serum provided by Pfeiffer to test his diagnostic reaction in vivo. Professor Gruber called out to me 'Durham! Kommen Sie her, schauen Sie an!' Before making our first injection with the mixtures of serum and vibrios, he had put a specimen under the microscope and there agglutination was displayed. A few days later, we had been making our mixtures in small sterilised glass pots, it happened that none were ready sterilised, so I had to make use of sterile test-tubes; those containing the mixture of culture and serum were left standing for a short time and then I called,' Herr Professor! Kommen Sie her, schauen Sie an!' the phenomenon of sedimentation was before his eyes! Thus there were two techniques available, the microscopic and the macroscopic.*

The discovery was quite unexpected and not anticipated by any hypothesis. It occurred incidentally in the course of another investigation, and macroscopic agglutination was found owing to the fortuitous lack of sterilised glass pots. [I am indebted to Professor H.R. Dean for showing me Durham's manuscript.]

Gowland Hopkins, whom many consider the father of biochemistry, gave his practical class a certain well-known test for proteins to carry out as an exercise, but all the students failed to elicit the reaction. Investigation revealed that the reaction was only obtained when the acetic acid employed contained an impurity, glyoxylic acid, which thereafter became the standard test reagent. Hopkins followed up this clue further and sought the group in the protein with which the glyoxylic acid reacted, and this led him to his famous isolation of tryptophane. [6]

When Weil and Felix were investigating cases of louse-borne typhus in Poland in 1915 they isolated the bacterium known as "Proteus X" from some patients. Thinking it might be the cause of the disease they tried agglutination of the organism with the patients' sera and obtained positive results. It was then found that Proteus X was not the causal organism of the disease; nevertheless agglutination of this organism proved to be a reliable and most valuable means of diagnosing typhus. In the course of their experimental study of this serological reaction Weil and Felix identified the O and H antigens and antibodies, and this discovery in turn opened up a completely new chapter in serology. Later it was found that in Malaya those cases of typhus contracted in the scrub failed to show agglutination to Proteus X19. Strangely enough a new strain of Proteus, obtained from England and believed to be a typical strain of Proteus X19, agglutinated with sera from cases of scrub typhus but not with sera from the cases contracted in the town (shop typhus), which were reacting satisfactorily with the Proteus X19 strain that had been used in many parts of the world. Later it transpired that scrub typhus and shop typhus were two different rickettsial diseases. How it came about that the strain of Proteus sent out from England was not only not typical Proteus X19, but had changed to just what was wanted to diagnose the other disease, remains a profound mystery [7].

Agglutination of red blood cells of the chick by influenza virus was first observed quite unexpectedly by Hirst and independently by McClelland and Hare when they were examining chick embryos infected with the virus. Fluid containing virus got mixed with blood cells which became agglutinated and the alert and observant scientists quickly followed up this clue. The discovery of this phenomenon has not only revolutionised much of our technique concerned with several viruses, but has opened up a method of approach to fundamental problems of virus-cell relationships [8,9].

Following this discovery, other workers tried haemagglutination with other viruses and Newcastle disease, fowl plague and vaccinia were found to produce the phenomenon. However it was again by chance observation that haemagglutination with the virus of mumps and later of mouse pneumonia was discovered.

Rickettsiae (microbes closely related to viruses) cause typhus and several other important diseases and are difficult to cultivate. Dr. Herald Cox spent much time and effort trying to improve on methods of growing them in tissue culture and had tried adding all sorts of extracts, vitamins and hormones without achieving anything. One day while setting up his tests he ran short of chick embryo tissue for tissue culture, so to make up the balance he used yolk sac which previously he, like everyone else, had discarded. When he later examined these cultures, to his "amazement and surprise," he found terrific numbers of the organisms in those tubes where he had happened to put yolk sac. A few nights later while in bed the idea occurred to him of inoculating the rickettsiae directly into the yolk sac of embryonated eggs. Getting out of bed at 4 a.m. he went to the laboratory and made the first inoculation of rickettsiae into the yolk sac. Thus was discovered an easy way of growing masses of rickettsiae, which has revolutionised the study of the many diseases they cause and made possible the production of effective vaccines against them. [Personal communication.]

Role of chance in discovery

These ten examples, together with nineteen others given in the Appendix and some of those in Chapters Four and Eight (Ed. Note: these chapter references are to Beveridge's book, not this one) provide striking illustration of the important part that chance plays in discovery. They are the more remarkable when one thinks of the failures and frustrations usually met in research. Probably the majority of discoveries in biology and medicine have been come upon unexpectedly, or at least had an element of chance in them, especially the most important and revolutionary ones. It is scarcely possible to foresee a discovery that breaks really new ground, because it is often not in accord with current beliefs. Frequently I have heard a colleague, relating some new finding, say almost apologetically, "I came across it by accident." Although it is common knowledge that sometimes chance is a factor in the making of a discovery, the magnitude of its importance is seldom realised and the significance of its role does not seem to have been fully appreciated or understood. Books have been written on scientific method omitting any reference to chance or empiricism in discovery.

Perhaps the most striking examples of empirical discoveries are to be found in chemotherapy where nearly all the great discoveries have been made by following a false hypothesis or a so-called chance observation. Elsewhere in this book are described the circumstances in which were discovered the therapeutic effects of quinine, salvarsan, sulphanilamide, diamidine, paraminobenzoic acid and penicillin. Subsequent rational research in each case provided only relatively small improvements. These facts are the more amazing when one thinks of the colossal amount of rational research that has been carried out in chemotherapy.

The research worker should take advantage of this knowledge of the importance of chance in discovery and not pass over it as an oddity or, worse, as something detracting from the credit due to the discoverer and therefore not to be dwelt upon. Although we cannot deliberately evoke that will-o'-the-wisp, chance, we can be on the alert for it, prepare ourselves to recognise it and profit by it when it comes. Merely realising the importance of chance may be of some help to the beginner. We need to train our powers of observation, to cultivate that attitude of mind of being constantly on the look-out for the unexpected and make a habit of examining every clue that chance presents. Discoveries are made by giving attention to the slightest clue. That aspect of the scientist's mind which demands convincing evidence should be reserved for the proof stage of the investigation. In research, an attitude of mind is required for discovery which is different from that required for proof, for discovery and proof are distinct processes. We should not be so obsessed with our hypothesis that we miss or neglect anything not directly bearing on it. With this in mind, Bernard insisted that, although hypotheses are essential in the planning of an experiment, once the experiment is commenced the observer should forget his hypothesis. People who are too fond of their hypotheses, he said, are not well fitted for making discoveries. The anecdote (related in Chapter Eight) about Bernard's work starting from the observation that the rabbits passed clear urine, provides a beautiful example of discovery involving chance, observation and a prepared mind.

A good maxim for the research man is "look out for the unexpected."

It is unwise to speak of luck in research as it may confuse our thinking. There can be no objection to the word when it is used to mean merely chance, but for many people luck is a metaphysical notion which in some mystical way influences events, and no such concept should be allowed to enter into scientific thinking. Nor is chance the only factor involved in these unexpected discoveries, as we shall discuss more fully in the next section. In the anecdotes cited, many of the opportunities might well have been passed over had not the workers been on the look-out for anything that might arise. The successful scientist gives attention to every unexpected happening or observation that chance offers and investigates those that seem to him promising. Sir Henry Dale has aptly spoken of opportunism in this connection. Scientists without the flair for discovery seldom notice or bother with the unexpected and so the occasional opportunity passes without them ever being aware of it. Alan Gregg wrote:

> *One wonders whether the rare ability to be completely attentive to, and to profit by, Nature's slightest deviation from the conduct expected of her is not the secret of the best research minds and one that explains why some men turn to most remarkably good advantage seemingly trivial accidents. Behind such attention lies an unremitting sensitivity.* [10]

Writing of Charles Darwin, his son said:

> *Everybody notices as a fact an exception when it is striking and frequent, but he had a special instinct for arresting an exception. A point apparently slight and unconnected with his present work is passed over by many a man almost unconsciously with some half considered explanation, which is in fact no explanation. It was just these things that he seized on to make a start from.* [11]

It is of the utmost importance that the role of chance be clearly understood. The history of discovery shows that chance plays an important part, but on the other hand it plays only one part even in those discoveries attributed to it. For this reason it is a misleading half-truth to refer to unexpected discoveries as "chance discoveries" or "accidental discoveries." If these discoveries were made by chance or accident alone, as many discoveries of this type would be made by any inexperienced scientist starting to dabble in research as by Bernard or Pasteur. The truth of the matter lies in Pasteur's famous saying: "In the field of observation, chance favours only the prepared mind." It is the interpretation of the chance observation which counts. The role of chance is merely to provide the opportunity and the scientist has to recognise it and grasp it.

Recognising chance opportunities

In reading of scientific discoveries one is sometimes struck by the simple and apparently easy observations which have given rise to great and far-reaching discoveries making scientists famous. But in retrospect we see the discovery with its significance established. Originally the discovery usually has no intrinsic significance; the discoverer gives it significance by relating it to other knowledge, and perhaps by using it to derive further knowledge. The difficulties in the way of making discoveries in which chance is involved may be discussed under the following headings.

(a) Infrequency of opportunities. Opportunities, in the form of significant clues, do not come very often. This is the only aspect affected by sheer chance, and even here the scientist does not play a purely passive role. The successful researchers are scientists who spend long hours working at the bench, and who do not confine their activities to the conventional but try out novel procedures, therefore they are exposed to the maximum extent to the risk of encountering a fortunate "accident."

(b) Noticing the clue. Acute powers of observation are often required to notice the clue, and especially the ability to remain alert and sensitive for the unexpected while watching for the expected. Noticing is discussed at length in the chapter on observation, and it need only be said here that it is mainly a mental process.

(c) Interpreting the clue. To interpret the clue and grasp its possible significance is the most difficult phase of all and requires the "prepared mind." Let us consider some instances of failure to grasp opportunities. The history of discovery teems with instances of lost opportunities—clues noticed but their significance not appreciated. Before Roentgen discovered X-rays, at least one other physicist had noticed evidence of the rays but was merely annoyed. Several people now recall having noticed the inhibition of staphylococcal colonies by moulds before Fleming followed it up to discover penicillin. Scott, for instance, reports that he saw it and considered it only a nuisance and he protests against the view that Fleming's discovery was due to chance, for, he says, it was due mainly to his perspicacity in seizing on the opportunity others had let pass [12]. Another interesting case is related by J.T. Edwards [13]. In 1919 he noticed that one of a group of cultures of *Brucella abortus* grew much more luxuriantly than the others and that it was contaminated with a mould. He called the attention of Sir John M'Fadyean to this, suggesting it might be of significance, but was greeted with scorn. It was not till later that it was discovered that *Br. abortus* grew much better in the presence of CO_2, which explains why Edwards' culture had grown much better in the presence of the mould. Bordet and others had casually noticed agglutination of bacteria by antisera, but none had seen the possibilities in it until Gruber and Durham did. Similarly, others had seen the phenomenon of bacteriophage lysis before Twort and d'Hérelle. F.M. Burnet for one now admits having seen agglutination of chick embryos' red blood cells in the presence of influenza virus and probably others had too but none followed it up till G.K. Hirst, and McClelland and Hare. Many bacteriologists had seen rough to smooth colony variation in bacteria before Arkwright investigated it and found it to be associated with change in virulence and antigenicity. It is now, of course, one of the fundamental facts in immunology and serology.

Sometimes the significance of the clue which chance brings our way is quite obvious, but at others it is just a trivial incident of significance only for the well prepared mind, the mind loaded with relevant data and ripe for discovery. When the mind has a lot of relevant but loosely connected data and vague ideas, a clarifying idea connecting them up may be helped to crystallise by some small incident. Just as a substance may crystallise out of solution in the presence of a nucleus consisting of a minute crystal with the correct configuration, so did the falling apple provide a model for Newton's mind. Sir Henry Souttar has pointed out that it is the content of the observer's brain, accumulated by years of work, that makes possible the moment of triumph. This aspect of chance observation will be discussed further in the chapters on observation and on intuition.

Anyone with an alertness of mind will encounter during the course of an investigation numerous interesting side issues that might be pursued. It is a physical impossibility to follow up all of these. The majority are not worth following, a few will reward investigation and the occasional one provides the opportunity of a lifetime. How to distinguish the promising clues is the very essence of the art of research. The scientist who has an independent mind and is able to judge the evidence on its merits rather than in light of prevailing conceptions is the one most likely to be able to realise the potentialities in something really new. He also needs imagination and a good fund of knowledge, to know whether or not his observation is new and to enable him to see the possible implications. In deciding whether a line of work should be followed, one should not be put off it merely because the idea has already been thought of by others or even been tried without it leading anywhere. This does not necessarily indicate that it is not good; many of the classic discoveries were anticipated in this way but were not properly developed until the right man came along. Edward Jenner was not the first to inoculate people with cowpox to protect them against smallpox, William Harvey was not the first to postulate circulation of the blood, Darwin was by no means the first to suggest evolution, Columbus was not the first European to go to America, Pasteur was not the first to propound the germ theory of disease, Lister was not the first to use carbolic acid as a wound antiseptic. But these men were the ones who fully developed these ideas and forced them on a reluctant world, and most credit rightly goes to them for bringing the discoveries to fruition. It is not only new ideas that lead to discoveries. Indeed few ideas are entirely original. Usually on close study of the origin of an idea, one finds that others had suggested it or something very like it previously. Charles Nicolle calls these early ideas that are not at first followed up, "precursor ideas." Exploiting opportunities

When a discovery has passed these hurdles and reached a stage where it is recognised and appreciated by its originator, there are still at least three more ways in which its general acceptance may be delayed.

(d) Failure to follow up the initial finding. The initial disclosure may not be made the most of because it may not be followed up and exploited. The most productive scientists have not been satisfied with clearing up the immediate question but having obtained some new knowledge, they made use of it to uncover something further and often of even greater importance. Steinhaeuser discovered in 1840 that cod-liver oil cured rickets but this enormously important fact remained unproved and no more than an opinion for the next eighty years [14]. In 1903 Theobald Smith discovered that some motile bacilli may exist in culture as the normal motile form or as a non-motile variant, and he demonstrated the significance of these two forms in immunological reactions [15]. This work passed almost unnoticed and was forgotten until the phenomenon was rediscovered in 1917 by Weil and Felix. It is now regarded as one of the fundamental facts in immunological reactions. Fleming described crude preparations of penicillin in 1929, but after a few years he dropped work on it without developing a therapeutic agent. He got no encouragement or assistance from others because they knew of many similar stories that had come to nothing. It was some years later that Florey took the work up from where Fleming left off and developed penicillin as a therapeutic agent.

(e) Lack of an application. There may be no possible applications of the discovery until years later. Neufeld discovered a rapid method of typing pneumococci in 1902, but it was not till 1931 that it became of any importance when type-specific serum therapy was introduced. Landsteiner discovered the human blood groups in 1901, but it was not till anticoagulants were found and blood transfusion was developed in the 1914-18 war that Landsteiner's discovery assumed importance and attracted attention.

(f) Indifference and opposition. Finally the discovery has to run the gauntlet of scepticism and often resistance on the part of Others. This can be one of the most difficult hurdles of all and it is here that the scientist occasionally has to fight and in the past has sometimes even lost his life. The psychology of mental resistance to new ideas, and actual opposition to discoveries are discussed in a later chapter.

Several of the points discussed in this and the preceding section may be illustrated by narrowing the story of Jenner's recognition of the potentialities of vaccination and his exploitation of it. Artificial immunisation against smallpox by means of inoculation with virulent smallpox material (variolation) had long been practised in the Orient. Some say that 1000 years B.C. it was the custom of China to insert material from smallpox lesions into the noses of children, others that variolation was introduced into China from India about A.D. 1000. Variolation was introduced from Constantinople into England about the middle of the eighteenth century and became an accepted though not very popular practice about the time that Edward Jenner was born. When Jenner was serving his apprenticeship between thirteen and eighteen years of age, his attention was called to the local belief in Gloucestershire that people who contracted cow-pox from cattle were subsequently immune to smallpox. Jenner found that the local physicians were mostly familiar with the traditional belief but did not take it seriously, although they also were encountering instances of failure of people to develop infection when given variolation after they had had cow-pox. Jenner evidently kept the matter in mind for years without doing anything about it. After returning to country practice he confided in a friend that he intended trying vaccination. He divulged his intentions under a bond of secrecy because he feared ridicule if they should fail. Meanwhile he was exercising his genius for taking pains and making accurate observation by carrying out experiments in other directions. He was making observations on the temperature and digestion of hibernating animals for John Hunter, experimenting with agricultural fertilisers for Joseph Banks and on his own behalf carrying out studies on how the young cuckoo gets rid of its fellow nestlings. He married at thirty-eight and when his wife had a child he inoculated him with swinepox and showed he was subsequently immune to smallpox. Still none of his colleagues—John Hunter among them—took much interest in Jenner's ideas about using cow-pox to vaccinate against smallpox and his first tentative paper on the subject was returned to him and apparently rejected. It was not till he was forty-seven years old (in the memorable year 1796) that he made his first successful vaccination from one human being to another. He transferred material from a pustule on the hand of a milkmaid, Sarah Nelmes, to an eight-year-old boy named James Phipps who thereby gained fame in the same odd way as did Joseph Meister for being the first person to receive Pasteur's treatment for rabies nearly a century later[1]. This is taken as the classical origin of vaccination but, as is often the case in the history of scientific discovery, the issue is not clear-cut. At least two others had actually performed it earlier but failed to follow it up. Jenner continued his experiments, and in 1798 published his famous *Inquiry*, reporting some twenty-three cases who were either vaccinated or had contracted cow-pox naturally and were subsequently shown to be immune to smallpox. Soon afterwards vaccination was taken up widely and spread throughout the world, despite severe opposition from certain quarters which curiously and interestingly enough persists even to-day in a fairly harmless form. Jenner suffered abuse but honours were soon showered on him from all quarters of the globe [16,17].

This history provides an admirable demonstration of how difficult it usually is to recognise the true significance of a new fact. Without knowing the full history one might well suppose Jenner's contribution to medical science a very simple one not meriting the fame subsequently bestowed on it. But neither John Hunter nor any of Jenner's colleagues and contemporaries were able to grasp the potentialities in advance, and similar opportunities had occurred and been let pass in other countries. There was an interval of thirty years after the experimentally minded Jenner himself became interested in the popular belief, before he performed the classical, crucial experiments. With our present conceptions of immunisation and of experimentation this may appear surprising

[1] Meister remained at the Pasteur Institute as concierge until the occupation of Paris by the Germans in 1940, when he committed suicide.

but we must remember how revolutionary the idea was, even given the fact that variolation was an accepted practice. The fact that others who had the same opportunity failed to discover vaccination and that it took Jenner thirty years shows what a difficult discovery it was to make. Animals were at that time regarded with repugnance by most people so the idea of infecting a human being with a disease of animals created utmost disgust. All sorts of dire results were prophesied, including "cow-mania" and "ox-faced children" (one was actually exhibited!). Like many great discoveries it did not require great erudition and it mainly devolved on having boldness and independence of mind to accept a revolutionary idea and imagination to realise its potentialities. But Jenner also had practical difficulties to overcome. He found that cows were subject to various sores on the teats, some of which also affected the milkers but did not give immunity to smallpox. Even present day virus specialists have great difficulty in distinguishing between the different types of sores that occur on cows' teats; and the position is complicated by observations suggesting that an attack of cow-pox does not confer immunity against a second attack of the same disease in the cow, a point Jenner himself noted.

Jenner's discovery has its element of irony which so often lends additional interest to scientific anecdotes. Modem investigators believe that the strains of vaccinia now used throughout the world for many years are not cow-pox but have derived from smallpox. Their origin is obscure but it seems that in the early days cow-pox and smallpox got mixed up and an attenuated strain of smallpox developed and was mistakenly used for cow-pox.

Summary

New knowledge very often has its origin in some quite unexpected observation or chance occurrence arising during an investigation. The importance of this factor in discovery should be fully appreciated and research workers ought deliberately to exploit it. Opportunities come more frequently to active bench workers and people who dabble in novel procedures. Interpreting the clue and realising its possible significance requires knowledge without fixed ideas, imagination, scientific taste, and a habit of contemplating all unexplained observations.

Appendix: Further examples of discoveries in which chance played a part

(1) It was not a physicist but a physiologist, Luigi Galvani, who discovered current electricity. He had dissected a frog and left it on a table near an electrical machine. When Galvani left it for a moment someone else touched the nerves of the leg with a scalpel and noticed this caused the leg muscles to contract. A third person noticed that the action was excited when there was a spark from the electric machine. When Galvani's attention was drawn to this strange phenomenon he excitedly investigated it and followed it up to discover current electricity. [19]

(2) In 1822 the Danish physicist, Oersted, at the end of a lecture happened to bring a wire, joined at its two extremities to a voltaic cell, to a position above and parallel to a magnetic needle. At first he had purposely held the wire perpendicular to the needle but nothing happened, but when by chance he held the wire horizontally and parallel to the needle he was astonished to see the needle change position. With quick insight he reversed the current and found that the needle deviated in the opposite direction. Thus by mere chance the relationship between electricity and magnetism was discovered and the path opened for the invention by Faraday of the electric dynamo. It was when telling of this that Pasteur made his famous remark: "In the field of observation chance favours only the prepared mind." Modem civilisation perhaps owes more to the discovery of electro-magnetic induction than to any other single discovery. [20]

(3) When von Roentgen discovered X-rays he was experimenting with electrical discharges in high vacua and using barium platinocyanide with the object of detecting invisible rays, but had no thought of such rays being able to penetrate opaque materials. Quite by chance he noticed that barium platinocyanide left on the bench near his vacuum tube became fluorescent although separated from the tube by black paper. He afterwards said: "I found by accident that the rays penetrated black paper." [21]

(4) When W.H. Perkin was only eighteen years old he tried to produce quinine by the oxidation of allyl-o-toluidine by potassium dichromate. He failed, but thought it might be interesting to see

what happened when a simpler base was treated with the same oxidiser. He chose aniline sulphate and thus produced the first aniline dye. But chance played an even bigger part than the bare facts indicate: had not his aniline contained as an impurity some p-toluidine the reaction could not have occurred. [21]

(5) During the first half of the nineteenth century it was firmly believed that animals were unable to manufacture carbohydrates, fats or proteins, all of which had to be obtained in the diet preformed from plants. All organic compounds were believed to be synthesised in plants whereas animals were thought to be capable only of breaking them down. Claude Bernard set out to investigate the metabolism of sugar and in particular to find where it is broken down. He fed a dog a diet rich in sugar and then examined the blood leaving the liver to see if the sugar had been broken down in the liver. He found a high sugar content, and then wisely carried out a similar estimation with a dog fed a sugar-free meal. To his astonishment he found also a high sugar content in the control animal's hepatic blood. He realised that contrary to all prevailing views the liver probably did produce sugar from something which is not sugar. Thereupon he set about an exhaustive series of experiments which firmly established the glycogenic activity of the liver. This discovery was due firstly to the fact that Bernard was meticulous in controlling every stage of his experiments, and secondly, to his ability to recognise the importance of a result discordant with prevailing ideas on the subject and to follow up the clue thus given. [22]

(6) A mixture of lime and copper sulphate was sprayed on posts supporting grape vines in Medoc with the object of frightening away pilferers. Millardet later noticed that leaves accidentally sprayed with the mixture were free from mildew. The following up of this clue led to the important discovery of the value of Bordeaux mixture in protecting fruit trees and vines from many diseases caused by fungi. [23]

(7) The property of formalin of removing the toxicity of toxins without affecting their antigenicity was discovered by Ramon by chance when he was adding antiseptics to filtrates with the object of preserving them. [24]

(8) The circumstances leading to the discovery of penicillin are widely known. Fleming was working with some plate cultures of staphylococci which he had occasion to open several times and, as often happens in such circumstances, they became contaminated. He noticed that the colonies of staphylococci around one particular colony died. Many bacteriologists would not have thought this particularly remarkable for it has long been known that some bacteria interfere with the growth of others. Fleming, however, saw the possible significance of the observation and followed it up to discover penicillin, although its development as a therapeutic agent was due to the subsequent work of Sir Howard Florey. The element of chance in this discovery is the more remarkable when one realises that that particular mould is not a very common one and, further, that subsequently a most extensive, world-wide search for other antibiotics has failed to date to discover anything else as good. It is of interest to note that the discovery would probably not have been made had not Fleming been working under "unfavourable" conditions in an old building where there was a lot of dust and contaminations were likely to occur." [25]

(9) J. Ungar [26] found that the action of penicillin on certain bacteria was slightly enhanced by the addition to the medium of paraminobenzoic acid (PABA). He did not explain what made him try this out but it seems likely that it was because PABA was known to be an essential growth factor for bacteria. Subsequently, Greiff, Pinkerton and Moragues [27] tested PABA to see if it enhanced the weak inhibitory effect which penicillin had against typhus rickettsiae. They found that PABA alone had a remarkably effective chemotherapeutic action against the typhus organisms. "This result was quite unexpected," they said. As a result of this work PABA became recognised as a valuable chemotherapeutic agent for the typhus group of fevers, against which previously nothing had been found effective. In the chapter on hypothesis I have described how salvarsan and sulphanilamide were discovered following an hypothesis that was not correct. Two other equally famous chemotherapeutic drugs were discovered only because they happened to be present as impurities in other substances which were being tested. Scientists closely associated with the work have told me the stories of these two discoveries but have

asked me not to publish them as other members of the team may not wish the way in which they made the discovery to be made public. Sir Lionel Whitby has told to me a story of a slightly different nature. He was conducting an experiment on the then new drug, sulphapyridine, and mice inoculated with pneumococci were being dosed throughout the day, but were not treated during the night. Sir Lionel had been out to a dinner party and before returning home visited the laboratory to see how the mice were getting on, and while there lightheartedly gave the mice a further dose of the drug. These mice resisted the pneumococci better than any mice had ever done before. Not till about a week later did Sir Lionel realise that it was the extra dose at midnight which had been responsible for the excellent results. From that time, both mice and men were dosed day and night when under sulphonamide treatment and they benefited much more than under the old routine.

(10) In my researches on foot-rot in sheep I made numerous attempts to prepare a medium in which the infective agent would grow. Reason led me to use sheep serum in the medium and the results were repeatedly negative. Finally I got a positive result and on looking back over my notes I saw that, in that batch of media, horse serum had been used in place of sheep serum because the supply of the latter had temporarily run out. With this clue it was a straightforward matter to isolate and demonstrate the causal agent of the disease—an organism which grows in the presence of horse serum but not sheep serum! Chance led to a discovery where reason had pointed in the opposite direction.

(11) The discovery that the human influenza virus is able to infect ferrets was a landmark in the study of human respiratory diseases. When an investigation on influenza was planned, ferrets were included among a long list of animals it was intended to try and infect sooner or later. However, some time before it was planned to try them, it was reported that a colony of ferrets was suffering from an illness which seemed to be the same as the influenza then affecting the people caring for them. Owing to this circumstantial evidence, ferrets were immediately tried and found susceptible to influenza. Afterwards it was found that the idea which prompted the tests in ferrets was quite mistaken for the dis-

ease occurring in the colony of ferrets was not influenza but distemper! [28]

(12) A group of English bacteriologists developed an effective method of sterilising air by means of a mist made from a solution of hexyl-resorcinol in propylene-glycol. They conducted a very extensive investigation trying out many mixtures. This one proved the best; the glycol was chosen merely as a suitable vehicle for the disinfectant, hexyl-resorcinol. Considerable interest was aroused by the work because of the possibility of preventing the spread of air-borne diseases by these means. When other investigators took up the work they found that the effectiveness of the mixture was due not to the hexyl-resorcinol but to the glycol. Subsequently, glycols proved to be some of the best substances for air disinfection. They were only introduced into this work as solvents for other supposedly more active disinfectants and were not at first suspected as having any appreciable disinfective action themselves." [29]

(13) Experiments were being conducted at Rothamsted Experimental Station on protecting plants from insects with various compounds, when it was noticed that those plants treated with boric acid were strikingly superior to the rest. Investigation by Davidson and Warington showed that the better growth had resulted because the plants required boron. Previously it had not been known that boron was of any importance in plant nutrition and even after this discovery, boron deficiency was for a time thought of as only of academic interest. Later, however, some diseases of considerable economic importance— "heart-rot" of sugar beet for example—were found to be manifestation of boron deficiency. [30]

(14) The discovery of selective weed-killers arose unexpectedly from studies on root nodule bacteria of clovers and plant growth stimulants. These beneficial bacterial nodules were found to exert their deforming action on the root hairs by secreting a certain substance. But when Nutman, Thornton and Quastel tested the action of this substance on various plants, they were surprised to find that it prevented germination and growth. Furthermore they found that this toxic effect was selective, being much greater against dicotyledon plants, which include most weeds, than

against monocotyledon plants, which include grain crops and grasses. They then tried related compounds and found some which are of great value in agriculture to-day as selective weed-killers. [31]

(15) Scientists working on the technicalities of food preservation tried prolonging the "life" of chilled meat by replacing the air by carbon dioxide which was known to have an inhibitory effect on the growth of microorganisms causing spoilage. Carbon dioxide, at the high concentration used, was found to cause an unpleasing discoloration of the meat and the whole idea was abandoned. Some time later, workers in the same laboratory were investigating a method of refrigeration which involved the release of carbon dioxide into the chamber in which the food was stored, and observations were carried out to see whether the gas had any undesirable effect. To their surprise the meat not only remained free from discoloration but even in the relatively low concentrations of carbon dioxide involved it kept in good condition much longer than ordinarily. From this observation was developed the important modem process of "gas storage" of meat in which 10-12 per cent carbon dioxide is used. At this concentration the gas effectively prolongs the "life" of chilled meat without causing discoloration. [32]

(16) I was investigating a disease of the genitalia of sheep known as balano-posthitis. It is a very long-lasting disease and was thought to be incurable except by radical surgery. Affected sheep were sent from the country to the laboratory for investigation but to my surprise they all healed spontaneously within a few days of arrival. At first it was thought that typical cases had not been sent, but further investigation showed that the self-imposed fasting of the sheep when placed in a strange environment had cured the disease. Thus it was found that this disease, refractory to other forms of treatment, could in most cases be cured by the simple expedient of fasting for a few days.

(17) Paul Ehrlich's discovery of the acid-fast method of staining tubercle bacilli arose from his having left some preparations on a stove which was later inadvertently lighted by someone. The heat of the stove was just what was required to make these waxy-coated bacteria take the stain. Robert Koch said "We owe it to this cir-

cumstance alone that it has become a general custom to search for the bacillus in sputum." [33]

(18) Dr. A.S. Parkes relates the following story of how he and his colleagues made the important discovery that the presence of glycerol enables living cells to be preserved for long periods at very low temperatures. "In the autumn of 1948 my colleagues. Dr. Audrey Smith and Mr. C. Polge, were attempting to repeat the results which Shaffner, Henderson and Card (1941) had obtained in the use of laevulose solutions to protect fowl spermatozoa against the effects of freezing and thawing. Small success attended the efforts, and pending inspiration a number of the solutions were put away in the cold-store. Some months later work was resumed with the same material and negative results were again obtained with all of the solutions except one which almost completely preserved motility in fowl spermatozoa frozen to -79 °C. This very curious result suggested that chemical changes in the laevulose, possibly caused or assisted by the flourishing growth of mould which had taken place during storage, had produced a substance with surprising powers of protecting living cells against the effects of freezing and thawing. Tests, however, showed that the mysterious solution not only contained no unusual sugars, but in fact contained no sugar at all. Meanwhile, further biological tests had shown that not only was motility preserved after freezing and thawing but, also, to some extent, fertilizing power. At this point, with some trepidation, the small amount (10-15 ml.) of the miraculous solution remaining was handed over to our colleague Dr. D. Elliott for chemical analysis. He reported that the solution contained glycerol, water, and a fair amount of protein! It was then realised that Mayer's albumen—the glycerol and albumen of the histologist—had been used in the course of morphological work on the spermatozoa at the same time as the laevulose solutions were being tested, and with them had been put away in the cold-store. Obviously there had been some confusion with the various bottles, though we never found out exactly what had happened. Tests with new material very soon showed that the albumen played no part in the protective effect, and our low temperature work became concentrated on the effects of glycerol in protecting living cells against the effects of low temperatures." [34]

(19) In a personal communication Dr. A.V. Nalbandov has given the following intriguing story of how he discovered the simple method of keeping experimental chickens alive after the surgical removal of the pituitary gland (hypophysectomy). "In 1940 I became interested in the effects of hypophysectomy of chickens. After I had mastered the surgical technique my birds continued to die and within a few weeks after the operation none remained alive. Neither replacement therapy nor any other precautions taken helped and I was about ready to agree with A.S. Parkes and R.T. Hill who had done similar operations in England, that hypophysectomized chickens simply cannot live. I resigned myself to doing a few short-term experiments and dropping the whole project when suddenly 98% of a group of hypophysectomized birds survived for 3 weeks and a great many lived for as long as 6 months. The only explanation I could find was that my surgical technique had improved with practice. At about this time, and when I was ready to start a long-term experiment, the birds again started dying and within a week both recently operated birds and those which had lived for several months, were dead. This, of course, argued against surgical proficiency. I continued with the project since I now knew that they could live under some circumstances which, however, eludes me completely. At about this time I had a second successful period during which mortality was very low. But, despite careful analysis of records (the possibility of disease and many other factors were considered and eliminated) no explanation was apparent. You can imagine how frustrating it was to be unable to take advantage of something that was obviously having a profound effect on the ability of these animals to withstand the operation. Late one night I was driving home from a party via a road which passes the laboratory. Even though it was 2 a.m. lights were burning in the animal rooms. I thought that a careless student had left them on so I stopped to turn them off. A few nights later I noted again that lights had been left on all night. Upon enquiry it turned out that a substitute janitor, whose job it was to make sure at midnight that all the windows were closed and doors locked, preferred to leave on the lights in the animal room in order to be able to find the exit door (the light switches not being near the door). Further checking showed that the two survival

periods coincided with the times when the substitute janitor was on the job. Controlled experiments soon showed that hypophysectomized chickens kept in darkness all died while chickens lighted for 2 one-hour periods nightly lived indefinitely. The explanation was that birds in the dark do not eat and develop hypoglycaemia from which they cannot recover, while birds which are lighted eat enough to prevent hypoglycaemia. Since that time we no longer experience any trouble in maintaining hypophysectomized birds for as long as we wish."

References

[1] Duclaux, E. *Pasteur: Histoire d'un Esprit.* (Paris: Sceaux, 1896).
[2] Gram, C. *Fortschritte der Medicirt,* Jakrg. II, p. 185 (1884).
[3] Cannon, W.B. *The Way of an Investigator.* (New York: W.W. Norton & Co. Inc., 1945). Permission to quote kindly granted by W.W. Norton & Co. Inc., New York, Publishers, who hold the copyright.
[4] Dunn, J. Shaw; Sheehan, H.L.; and McLetchie, N.G.B. *Lancet,* Vol. 1, p. 484 (1943).
[5] Dale, H.H. "Accident and Opportunism in Medical Research" *British Medical Journal,* Sept. 4th, p. 451 (1948).
[6] Stephenson, M. "F. Gowland Hopkins." *Biochemical Journal,* Vol. 42, 161 (1948).
[7] Felix, A. Personal Communication.
[8] Hirst, G.K. *Science,* Vol. 94, 22 (1941).
[9] McClelland, L., & Hare, R., *Canadian Public Health Journal,* Vol. 32, 530 (1941).
[10] Gregg, A. *The Furtherance of Medical Research.* (London: Oxford University Press; New Haven: Yale University Press, 1941) Permission to quote kindly granted by Oxford University Press.
[11] Darwin, F. *Life and Letters of C. Darwin.*(London: John Murray, 1888).
[12] Scott, W.M. *Veterinary Record,* Vol. 59, 680 (1947).
[13] Edwards, J.T. *Veterinary Record,* Vol. 60, 44 (1948).
[14] Trotter, W. *Collected Papers of Wilfred Trotter.* (London: Oxford University Press, 1941). Permission to quote kindly granted by Oxford University Press, London.

[15] Topley, W.W.C. & Wilson, G.S. *The Principles of Bacteriology and Immunity.* (London: Edward Arnold & Co., 1929).
[16] Bashford, H.H. *The Harley Street Calendar.* (London: Constable & Co. Ltd., 1929)
[17] Robinson, V. *Pathfinders in Medicine.* (New York: Medical Life Press, 1929).
[18] Zinsser, H. *As I Remember Him.* (London: Macmillan & Co., Ltd; Boston: Brown & Co.; Atlantic Monthly Press 1940). Permission to quote kindly granted by the publishers.
[19] Dubos, R.J. *Louis Pasteur: Free Lance of Science.* (Boston: Little, Brown & Co., 1950).
[20] Pearce, R.M. In *Medical Research and Education.* (New York: The Science Press, 1913).
[21] Baker, J.R. *The Scientific Life.* (London: George Allen & Unwin Ltd., 1942).
[22] Foster, M. *Claude Bernard.* (London: T. Fisher Unwin Ltd., 1899). Permission to quote kindly granted by T. Fisher Unwin Ltd., London. [23] Snedecor, G.W. *Statistical Methods applied to Experiments in Agriculture and Biology.* (Ames, IA: Collegiate Press Inc., 1938).
[24] Nicolle, C. *Biologic de l'Invention.* (Paris: Alcan, 1932).
[25] Fleming, A. *Nature*, Vol. 155, 796 (1945).
[26] Ungar, J. *Nature*, Vol. 152, 245 (1943).
[27] Greiff, D., Pinkerton, H., & Moragues, V., *Journal of Experimental Medicine*, Vol. 80,561 (1944).
[28] Andrewes, C.H. (1948). Personal communication.
[29] Robertson, O.H., Bigg, E., Puck, T.T., & Miller, B.F. *Journal of Experimental Medicine*, Vol. 75, 593 (1942).
[30] Warington, K. *Annals of Botany*, Vol. 37, 629 (1923).
[31] Nutman, P.S., Thornton, H.G., and Quastel, J.H. *Nature*, Vol. 155, 498 (1945).
[32] Bate-Smith, E.C. (1948). Personal Communication.
[33] Marquardt, M. *Paul Ehrlich.* (London: Wm. Heinemann Ltd., 1949).
[34] Parkes, A.S. *Proceedings of the III International Congress on Animal Reproduction*, Cambridge, 25-30 June, 1956.

Part IV
Life Lessons for Research

16

You and Your Research
Richard Hamming

Richard Hamming (February 11, 1915 – January 7, 1998)—was a famous mathematician and computer science at Bell Laboratories in Murray Hill, New Jersey. During his long and productive career, Hamming was one of the first scientist to use the new information theory of Claude Shannon for practical advances in telecommunications. The created one of the first error correcting codes which are often used in computer memory in contrast to the error detecting, but not correcting, bit parity codes in telecommunications. He was also a founder of the Association for Computing Machinery (ACM) which is one of the biggest societies and journal publishers in computer science today. In this talk, he gives sometimes counterintuitive advice on how to be productive in research and which pitfalls to avoid.

At a seminar in the Bell Communications Research Colloquia Series, Dr. Richard W. Hamming, a Professor at the Naval Postgraduate School in Monterey, California and a retired Bell Labs scientist, gave a very interesting and stimulating talk, 'You and Your Research' to an overflow audience of some 200 Bellcore staff members and visitors at the Morris Research and Engineering Center on March 7, 1986. This talk centered on Hamming's observations and research on the question "Why do so few scientists make significant contributions and so many are forgotten in the long run?" From his more than forty years of experience, thirty of which were at Bell Laboratories, he has made a number of direct observations, asked very pointed questions of scientists about what, how, and why they did things, studied the lives of great scientists and great contributions, and has done introspection and studied theories of creativity. The talk is about what he has

---

* Reprinted from Richard Hamming You and your research. In: Ed. J.F. Kaiser *Transcription of the Bell Communications Research Colloquium Seminar*; 7 March 1986; Morristown, New Jersey, United States.

learned in terms of the properties of the individual scientists, their abilities, traits, working habits, attitudes, and philosophy.

In order to make the information in the talk more widely available, the tape recording that was made of that talk was carefully transcribed. This transcription includes the discussions which followed in the question and answer period. As with any talk, the transcribed version suffers from translation as all the inflections of voice and the gestures of the speaker are lost; one must listen to the tape recording to recapture that part of the presentation. While the recording of Richard Hamming's talk was completely intelligible, that of some of the questioner's remarks were not. Where the tape recording was not intelligible I have added in parentheses my impression of the questioner's remarks. Where there was a question and I could identify the questioner, I have checked with each to ensure the accuracy of my interpretation of their remarks.

Introduction of Dr. Richard W. Hamming

As a speaker in the Bell Communications Research Colloquium Series, Dr. Richard W. Hamming of the Naval Postgraduate School in Monterey, California, was introduced by Alan G. Chynoweth, Vice President, Applied Research, Bell Communications Research.

Alan G. Chynoweth: Greetings colleagues, and also to many of our former colleagues from Bell Labs who, I understand, are here to be with us today on what I regard as a particularly felicitous occasion. It gives me very great pleasure indeed to introduce to you my old friend and colleague from many many years back, Richard Hamming, or Dick Hamming as he has always been known to all of us.

Dick is one of the all time greats in the mathematics and computer science arenas, as I'm sure the audience here does not need reminding. He received his early education at the Universities of Chicago and Nebraska, and got his Ph.D. at Illinois; he then joined the Los Alamos project during the war. Afterwards, in 1946, he joined Bell Labs. And that is, of course, where I met Dick—when I joined Bell Labs in their physics research organization. In those days, we were in the habit of lunching together as a physics group, and for some reason this strange fellow from mathematics was always pleased to join us. We were always happy to have him with us because he brought so many unorthodox ideas and views. Those lunches were stimulating, I can assure you.

While our professional paths have not been very close over the years, nevertheless I've always recognized Dick in the halls of Bell Labs and have always had tremendous admiration for what he was doing. I think the record speaks for itself. It is too long to go through all the details, but let me point out, for example, that he has written seven books and of those seven books which tell of various areas of mathematics and computers and coding and information theory, three are already well into their second edition. That is testimony indeed to the prolific output and the stature of Dick Hamming.

I think I last met him—it must have been about ten years ago—at a rather curious little conference in Dublin, Ireland where we were both speakers. As always, he was tremendously entertaining. Just one more example of the provocative thoughts that he comes up with: I remember him saying, "There are wavelengths that people cannot see, there are sounds that people cannot hear, and maybe computers have thoughts that people cannot think." Well, with Dick Hamming around, we don't need a computer. I think that we are in for an extremely entertaining talk.

The Talk: "You and Your Research" by Dr. Richard W. Hamming

It's a pleasure to be here. I doubt if I can live up to the Introduction. The title of my talk is, "You and Your Research." It is not about managing research, it is about how you individually do your research. I could give a talk on the other subject—but it's not, it's about you. I'm not talking about ordinary run-of-the-mill research; I'm talking about great research. And for the sake of describing great research I'll occasionally say Nobel-Prize type of work. It doesn't have to gain the Nobel Prize, but I mean those kinds of things which we perceive are significant things. Relativity, if you want, Shannon's information theory, any number of outstanding theories—that's the kind of thing I'm talking about.

Now, how did I come to do this study? At Los Alamos I was brought in to run the computing machines which other people had got going, so those scientists and physicists could get back to business. I saw I was a stooge. I saw that although physically I was the same, they were different. And to put the thing bluntly, I was envious. I wanted to know why they were so different from me. I saw Feynman up close. I saw Fermi and Teller. I saw Oppenheimer. I saw Hans Bethe: he was my boss. I saw quite a few very capable people. I became very interested in the difference between those who do and those who might have done.

When I came to Bell Labs, I came into a very productive department. Bode was the department head at the time; Shannon was there, and there were other people. I continued examining the questions, "Why?" and "What is the difference?" I continued subsequently by reading biographies, autobiographies, asking people questions such as: "How did you come to do this?" I tried to find out what are the differences. And that's what this talk is about.

Now, why is this talk important? I think it is important because, as far as I know, each of you has one life to live. Even if you believe in reincarnation it doesn't do you any good from one life to the next! Why shouldn't you do significant things in this one life, however you define significant? I'm not going to define it—you know what I mean. I will talk mainly about science because that is what I have studied. But so far as I know, and I've been told by others, much of what I say applies to many fields. Outstanding work is characterized very much the same way in most fields, but I will confine myself to science.

In order to get at you individually, I must talk in the first person. I have to get you to drop modesty and say to yourself, "Yes, I would like to do first-class work." Our society frowns on people who set out to do really good work. You're not supposed to; luck is supposed to descend on you and you do great things by chance. Well, that's a kind of dumb thing to say. I say, why shouldn't you set out to do something significant. You don't have to tell other people, but shouldn't you say to yourself, "Yes, I would like to do something significant."

In order to get to the second stage, I have to drop modesty and talk in the first person about what I've seen, what I've done, and what I've heard. I'm going to talk about people, some of whom you know, and I trust that when we leave, you won't quote me as saying some of the things I said.

Let me start not logically, but psychologically. I find that the major objection is that people think great science is done by luck. It's all a matter of luck. Well, consider Einstein. Note how many different things he did that were good. Was it all luck? Wasn't it a little too repetitive? Consider Shannon. He didn't do just information theory. Several years before, he did some other good things and some which are still locked up in the security of cryptography. He did many good things.

You see again and again, that it is more than one thing from a good person. Once in a while a person does only one thing in his whole life, and we'll talk about that later, but a lot of times there is repetition. I claim that luck will not cover everything. And I will cite Pasteur who said, "Luck favors the prepared mind." And I think that says it the way I believe it. There is indeed an element of luck, and no, there isn't. The prepared mind sooner or later finds something important and does it. So yes, it is luck. The particular thing you do is luck, but that you do something is not.

For example, when I came to Bell Labs, I shared an office for a while with Shannon. At the same time he was doing information theory, I was doing coding theory. It is suspicious that the two of us did it at the same place and at the same time—it was in the atmosphere. And you can say, "Yes, it was luck." On the other hand you can say, "But why of all the people in Bell Labs then were those the two who did it?" Yes, it is partly luck, and partly it is the prepared mind; but 'partly' is the other thing I'm going to talk about. So, although I'll come back several more times to luck, I want to dispose of this matter of luck as being the sole criterion whether you do great work or not. I claim you have some, but not total, control over it. And I will quote, finally, Newton on the matter. Newton said, "If others would think as hard as I did, then they would get similar results."

One of the characteristics you see, and many people have it including great scientists, is that usually when they were young they had independent thoughts and had the courage to pursue them. For example, Einstein, somewhere around 12 or 14, asked himself the question, "What would a light wave look like if I went with the velocity of light to look at it?" Now he knew that electromagnetic theory says you cannot have a stationary local maximum. But if he moved along with the velocity of light, he would see a local maximum. He could see a contradiction at the age of 12, 14, or somewhere around there, that everything was not right and that the velocity of light had something peculiar. Is it luck that he finally created special relativity? Early on, he had laid down some of the pieces by thinking of the fragments. Now that's the necessary but not sufficient condition. All of these items I will talk about are both luck and not luck.

How about having lots of 'brains?' It sounds good. Most of you in this room probably have more than enough brains to do first-class work. But great work is something else than mere brains. Brains are measured in various ways. In mathematics, theoretical physics, astrophysics, typically brains correlates to a great extent with the ability to manipulate symbols. And so the typical I.Q. test is apt to score them fairly high. On the other hand, in other fields it is something different. For example, Bill Pfann, the fellow who did zone melting, came into my office one day. He had this idea dimly in his mind about what he wanted and he had some equations. It was pretty clear to me that this man didn't know much mathematics and he wasn't really articulate. His problem seemed interesting so I took it home and did a little work. I finally showed him how to run computers so he could compute his own answers. I gave him the power to compute. He went ahead, with negligible recognition from his own department, but ultimately he has collected all the prizes in the field. Once he got well started, his shyness, his awkwardness, his inarticulateness, fell away and he became much more productive in many other ways. Certainly he became much more articulate.

And I can cite another person in the same way. I trust he isn't in the audience, i.e. a fellow named Clogston. I met him when I was working on a problem with John Pierce's group and I didn't think he had much. I asked my friends who had been with him at school, "Was he like that in graduate school?" "Yes," they replied. Well I would have fired the fellow, but J.R. Pierce was smart and kept him on. Clogston finally did the Clogston cable. After that there was a steady stream of good ideas. One success brought him confidence and courage.

One of the characteristics of successful scientists is having courage. Once you get your courage up and believe that you can do important problems, then you can. If you think you can't, almost surely you are not going to. Courage is one of the things that Shannon had supremely. You have only to think of his major theorem. He wants to create a method of coding, but he doesn't know what to do so he makes a random code. Then he is stuck. And then he asks the impossible question, "What would the average random code do?" He then proves that the average code is arbitrarily good, and that therefore there must be at least one good code. Who but a man of infinite courage could have dared to think those thoughts? That is the characteristic of great scientists; they have courage. They will go forward under incredible circumstances; they think and continue to think.

Age is another factor which the physicists particularly worry about. They always are saying that you have got to do it when you are young or you will never do it. Einstein did things very early, and all the quantum mechanics fellows were disgustingly young when they did their best work. Most mathematicians, theoretical physicists, and astrophysicists do what we consider their best work when they are young. It is not that they don't do good work in their old age but what we value most is often what they did early. On the other hand, in music, politics and literature, often what we consider their best work was done late. I don't know how whatever field you are in fits this scale, but age has some effect.

But let me say why age seems to have the effect it does. In the first place if you do some good work you will find yourself on all kinds of committees and unable to do any more work. You may find yourself as I saw Brattain when he got a Nobel Prize. The day the prize was announced we all assembled in Arnold Auditorium; all three winners got up and made speeches. The third one, Brattain, practically with tears in his eyes, said, "I know about this Nobel-Prize effect and I am not going to let it affect me; I am going to remain good old Walter Brattain." Well I said to myself, "That is nice." But in a few weeks I saw it was affecting him. Now he could only work on great problems.

When you are famous it is hard to work on small problems. This is what did Shannon in. After information theory, what do you do for an encore? The great scientists often make this error. They fail to continue to plant the little acorns from which the mighty oak trees grow. They try to get the big thing right off. And that isn't the way things go. So that is another reason why you find that when you get early recognition it seems to sterilize you. In fact I will give you my favorite quotation of many years. The Institute for Advanced Study in Princeton, in my opinion, has ruined more good scientists than any institution has created, judged by what they did before they came and judged by what they did after. Not that they weren't good afterwards, but they were superb before they got there and were only good afterwards.

This brings up the subject, out of order perhaps, of working conditions. What most people think are the best working conditions, are not. Very clearly they are not because people are often most productive when working conditions are bad. One of the better times of the Cambridge Physical Laboratories was when they had practically shacks—they did some of the best physics ever.

I give you a story from my own private life. Early on it became evident to me that Bell Laboratories was not going to give me the conventional acre of programming people to program computing machines in absolute binary. It was clear they weren't going to. But that was the way everybody did it. I could go to the West Coast and get a job with the airplane companies without any trouble, but the exciting people were at Bell Labs and the fellows out there in the airplane companies were not. I thought for a long while about, "Did I want to go or not?" and I wondered how I could get the best of two possible worlds. I finally said to myself, "Hamming, you think the machines can do practically everything. Why can't you make them write programs?" What appeared at first to me as a defect forced me into automatic programming very early. What appears to be a fault, often, by a change of viewpoint, turns out to be one of the greatest assets you can have. But you are not likely to think that when you first look the thing and say, "Gee, I'm never going to get enough programmers, so how can I ever do any great programming?"

And there are many other stories of the same kind; Grace Hopper has similar ones. I think that if you look carefully you will see that often the great scientists, by turning the problem around a bit, changed a defect to an asset. For example, many scientists when they found they couldn't do a problem finally began to study why not. They then turned it around the other way and said, "But of course, this is what it is" and got an important result. So ideal working conditions are very strange. The ones you want aren't always the best ones for you.

Now for the matter of drive. You observe that most great scientists have tremendous drive. I worked for ten years with John Tukey at Bell Labs. He had tremendous drive. One day about three or four years after I joined, I discovered that John Tukey was slightly younger than I was. John was a genius and I clearly was not. Well I went storming into Bode's office and said, "How can anybody my age know as much as John Tukey does?" He leaned back in his chair, put his hands behind his head, grinned slightly, and said, "You would be surprised Hamming, how much you would know if you worked as hard as he did that many years." I simply slunk out of the office!

What Bode was saying was this: "Knowledge and productivity are like compound interest." Given two people of approximately the same ability and one person who works ten percent more than the other, the latter will more than twice outproduce the former. The more you know, the more you learn; the more you learn, the more you can do; the more you can do, the more the opportunity—it is very much like compound interest. I don't want to give you a rate, but it is a very high rate. Given two people with exactly the same ability, the one person who manages day in and day out to get in one more hour of thinking will be tremendously more productive over a lifetime. I took Bode's remark to heart; I spent a good deal more of my time for some years trying to work a bit harder and I found, in fact, I could get more work done. I don't like to say it in front of my wife, but I did sort of neglect her sometimes; I needed to study. You have to neglect things if you intend to get what you want done. There's no question about this.

On this matter of drive Edison says, "Genius is 99% perspiration and 1% inspiration." He may have been exaggerating, but the idea is that solid work, steadily applied, gets you surprisingly far. The steady application of effort with a little bit more work, intelligently applied is what does it. That's the trouble; drive, misapplied, doesn't get you anywhere. I've often wondered why so many of my good friends at Bell Labs who worked as hard or harder than I did, didn't have so much to show for it. The misapplication of effort is a very serious matter. Just hard work is not enough—it must be applied sensibly.

There's another trait on the side which I want to talk about; that trait is ambiguity. It took me a while to discover its importance. Most people like to believe something is or is not true. Great scientists tolerate ambiguity very well. They believe the theory enough to go ahead; they doubt it enough to notice the errors and faults so they can step forward and create the new replacement theory. If you believe too much you'll never notice the flaws; if you doubt too much you won't get started. It requires a lovely balance. But most great scientists are well aware of why their theories are true and they are also well aware of some slight misfits which don't quite fit and they don't forget it. Darwin writes in his autobiography that he found it necessary to write down every piece of evidence which appeared to contradict his beliefs because otherwise they would disappear from his mind. When you find apparent flaws you've got to be sensitive and keep track of those things, and keep an eye out for how they can be explained or how the theory can be changed to fit them. Those are often the great contributions. Great contributions are rarely done by adding another decimal place. It comes down to an emotional commitment. Most great scientists are completely committed to their problem. Those who don't become committed seldom produce outstanding, first-class work.

Now again, emotional commitment is not enough. It is a necessary condition apparently. And I think I can tell you the reason why. Everybody who has studied creativity is driven finally to saying, "creativity comes out of your subconscious." Somehow, suddenly, there it is. It just appears. Well, we know very little about the subconscious; but one thing you are pretty well aware of is that your dreams also come out of your subconscious. And you're aware your dreams are, to a fair extent, a reworking of the experiences of the day. If you are deeply immersed and committed to a topic, day after day after day, your subconscious has nothing to do but work on your problem. And so you wake up one morning, or on some afternoon, and there's the answer. For those who don't get committed to their current problem, the subconscious goofs off on other things and doesn't produce the big result. So the way to manage yourself is that when you have a real important problem you don't let anything else get the center of your attention—you keep your thoughts on the problem. Keep your subconscious starved so it has to work on *your* problem, so you can sleep peacefully and get the answer in the morning, free.

Now Alan Chynoweth mentioned that I used to eat at the physics table. I had been eating with the mathematicians and I found out that I already knew a fair amount of mathematics; in fact, I wasn't learning much. The physics table was, as he said, an exciting place, but I think he exaggerated on how much I contributed. It was very interesting to listen to Shockley, Brattain, Bardeen, J.B. Johnson, Ken McKay and other people, and I was learning a lot. But unfortunately a Nobel Prize came, and a promotion came, and what was left was the dregs. Nobody wanted what was left. Well, there was no use eating with them!

Over on the other side of the dining hall was a chemistry table. I had worked with one of the fellows, Dave McCall; furthermore he was courting our secretary at the time. I went over and said, "Do you mind if I join you?" They can't say no, so I started eating with them for a while. And I started asking, "What are the important problems of your field?" And after a week or so, "What important problems are you working on?" And after some more time I came in one day and said, "If what you are doing is not important, and if you don't think it is going to lead to something important, why are you at Bell Labs working on it?" I wasn't welcomed after that; I had to find somebody else to eat with! That was in the spring.

In the fall, Dave McCall stopped me in the hall and said, "Hamming, that remark of yours got underneath my skin. I thought about it all summer, i.e. what were the important problems in my field. I haven't changed my research," he says, "but I think it was well worthwhile." And I said, "Thank you Dave," and went on. I noticed a couple of months later he was made the head of the department. I noticed the other day he was a Member of the National Academy of Engineering. I noticed he has succeeded. I have never heard the names of any of the other fellows at that table mentioned in science and scientific circles. They were unable to ask themselves, "What are the important problems in my field?"

If you do not work on an important problem, it's unlikely you'll do important work. It's perfectly obvious. Great scientists have thought through, in a careful way, a number of important problems in their field, and they keep an eye on wondering how to attack them. Let me warn you, 'important problem' must be phrased carefully. The three outstanding problems in physics, in a certain sense, were never worked on while I was at Bell Labs. By important I mean guaranteed a Nobel Prize and any sum of money you want to mention. We didn't work on (1) time travel, (2) teleportation, and (3) antigravity. They are not important problems because we do not have an attack. It's not the consequence that makes a problem important, it is that you have a reasonable attack. That is what makes a problem important. When I say that most scientists don't work on important problems, I mean it in that sense. The average scientist, so far as I can make out, spends almost all his time working on problems which he believes will not be important and he also doesn't believe that they will lead to important problems.

I spoke earlier about planting acorns so that oaks will grow. You can't always know exactly where to be, but you can keep active in places where something might happen. And even if you believe that great science is a matter of luck, you can stand on a mountain top where lightning strikes; you don't have to hide in the valley where you're safe. But the average scientist does routine safe work almost all the time and so he (or she) doesn't produce much. It's that simple. If you want to do great work, you clearly must work on important problems, and you should have an idea.

Along those lines at some urging from John Tukey and others, I finally adopted what I called "Great Thoughts Time." When I went to lunch Friday noon, I would only discuss great thoughts after that. By great thoughts I mean ones like: "What will be the role of computers in all of AT&T?," "How will computers change science?" For example, I came up with the observation at that time that nine out of ten experiments were done in the lab and one in ten on the computer. I made a remark to the vice presidents one time, that it would be reversed, i.e. nine out of ten experiments would be done on the computer and one in ten in the lab. They knew I was a crazy mathematician and had no sense of reality. I knew they were wrong and they've been proved wrong while I have been proved right. They built laboratories when they didn't need them. I saw that computers were transforming science because I spent a lot of time asking "What will be the impact of computers on science and how can I change it?" I asked myself, "How is it going to change Bell Labs?" I remarked one time, in the same address, that more than one-half of the people at Bell Labs will be interacting closely with computing machines before I leave. Well, you all have terminals now. I thought hard about where was my field going, where were the opportunities, and what were the important things to do. Let me go there so there is a chance I can do important things.

Most great scientists know many important problems. They have something between 10 and 20 important problems for which they are looking for an attack. And when they see a new idea come up, one hears them say "Well that bears on this problem." They drop all the other things and get after it. Now I can tell you a horror story that was told to me but I can't vouch for the truth of it. I was sitting in an airport talking to a friend of mine from Los Alamos about how it was lucky that the fission experiment occurred over in Europe when it did because that got us working on the atomic bomb here in the US. He said "No; at Berkeley we had gathered a bunch of data; we didn't get around to reducing it because we were building some more equipment, but if we had reduced that data we would have found fission." They had it in their hands and they didn't pursue it. They came in second!

The great scientists, when an opportunity opens up, get after it and they pursue it. They drop all other things. They get rid of other things and they get after an idea because they had already thought the thing through. Their minds are prepared; they see the opportunity and they go after it. Now of course lots of times it doesn't work out, but you don't have to hit many of them to do some great science. It's kind of easy. One of the chief tricks is to live a long time!

Another trait, it took me a while to notice. I noticed the following facts about people who work with the door open or the door closed. I notice that if you have the door to your office closed, you get more work done today and tomorrow, and you are more productive than most. But 10 years later somehow you don't know quite know what problems are worth working on; all the hard work you do is sort of tangential in importance. He who works with the door open gets all kinds of interruptions, but he also occasionally gets clues as to what the world is and what might be important. Now I cannot prove the cause and effect sequence because you might say, "The closed door is symbolic of a closed mind." I don't know. But I can say there is a pretty good correlation between those who work with the doors open and those who ultimately do important things, although people who work with doors closed often work harder. Somehow they seem to work on slightly the wrong thing—not much, but enough that they miss fame.

I want to talk on another topic. It is based on the song which I think many of you know, "It ain't what you do, it's the way that you do it." I'll start with an example of my own. I was conned into doing on a digital computer, in the absolute binary days, a problem which the best analog computers couldn't do. And I was getting an answer. When I thought carefully and said to myself, "You know, Hamming, you're going to have to file a report on this military job; after you spend a lot of money you're going to have to account for it and every analog installation is going to want the report to see if they can't find flaws in it." I was doing the required integration by a rather crummy method, to say the least, but I was getting the answer. And I realized that in truth the problem was not just to get the answer; it was to demonstrate for the first time, and beyond question, that I could beat the analog computer on its own ground with a digital machine. I reworked the method of solution, created a theory which was nice and elegant, and changed the way we computed the answer; the results were no different. The published report had an elegant method which was later known for years as "Hamming's Method of Integrating Differential Equations." It is somewhat obsolete now, but for a while it was a very good method. By changing the problem slightly, I did important work rather than trivial work.

In the same way, when using the machine up in the attic in the early days, I was solving one problem after another after another; a fair number were successful and there were a few failures. I went home one Friday after finishing a problem, and curiously enough I wasn't happy; I was depressed. I could see life being a long sequence of one problem after another after another. After quite a while of thinking I decided, "No, I should be in the mass production of a variable product. I should be concerned with *all* of next year's problems, not just the one in front of my face." By changing the question I still got the same kind of results or better, but I changed things and did important work. I attacked the major problem—How do I conquer machines and do all of next year's problems when I don't know what they are going to be? How do I prepare for it? How do I do this one so I'll be on top of it? How do I obey Newton's rule? He said, "If I have seen further than others, it is because I've stood on the shoulders of giants." These days we stand on each other's feet!

You should do your job in such a fashion that others can build on top of it, so they will indeed say, "Yes, I've stood on so and so's shoulders and I saw further." The essence of science is cumulative. By changing a problem slightly you can often do great work rather than merely good work. Instead of attacking isolated problems, I made the resolution that I would never again solve an isolated problem except as characteristic of a class.

Now if you are much of a mathematician you know that the effort to generalize often means that the solution is simple. Often by stopping and saying, "This is the problem he wants but this is characteristic of so and so. Yes, I can attack the whole class with a far superior method than the particular one because I was earlier embedded in needless detail." The business of abstraction frequently makes things simple. Furthermore, I filed away the methods and prepared for the future problems.

To end this part, I'll remind you, "It is a poor workman who blames his tools—the good man gets on with the job, given what he's got, and gets the best answer he can." And I suggest that by altering the problem, by looking at the thing differently, you can make a great deal of difference in your final productivity because you can either do it in such a fashion that people can indeed build on what you've done, or you can do it in such a fashion that the next person has to essentially duplicate again what you've done. It isn't just a matter of the job, it's the way you write the report, the way you write the paper, the whole attitude. It's just as easy to do a broad, general job as one very special case. And it's much more satisfying and rewarding!

I have now come down to a topic which is very distasteful; it is not sufficient to do a job, you have to sell it. 'Selling' to a scientist is an awkward thing to do. It's very ugly; you shouldn't have to do it. The world is supposed to be waiting, and when you do something great, they should rush out and welcome it. But the fact is everyone is busy with their own work. You must present it so well that they will set aside what they are doing, look at what you've done, read it, and come back and say, "Yes, that was good." I suggest that when you open a journal, as you turn the pages, you ask why you read some articles and not others. You had better write your report so when it is published in the Physical Review, or wherever else you want it, as the readers are turning the pages they won't just turn your pages but they will stop and read yours. If they don't stop and read it, you won't get credit.

There are three things you have to do in selling. You have to learn to write clearly and well so that people will read it, you must learn to give reasonably formal talks, and you also must learn to give informal talks. We had a lot of so-called 'back room scientists.' In a conference, they would keep quiet. Three weeks later after a decision was made they filed a report saying why you should do so and so. Well, it was too late. They would not stand up right in the middle of a hot conference, in the middle of activity, and say, "We should do this for these reasons." You need to master that form of communication as well as prepared speeches.

When I first started, I got practically physically ill while giving a speech, and I was very, very nervous. I realized I either had to learn to give speeches smoothly or I would essentially partially cripple my whole career. The first time IBM asked me to give a speech in New York one evening, I decided I was going to give a really good speech, a speech that was wanted, not a technical one but a broad one, and at the end if they liked it, I'd quietly say, "Any time you want one I'll come in and give you one." As a result, I got a great deal of practice giving speeches to a limited audience and I got over being afraid. Furthermore, I could also then study what methods were effective and what were ineffective.

While going to meetings I had already been studying why some papers are remembered and most are not. The technical person wants to give a highly limited technical talk. Most of the time the audience wants a broad general talk and wants much more survey and background than the speaker is willing to give. As a result, many talks are ineffective. The speaker names a topic and suddenly plunges into the details he's solved. Few people in the audience may follow. You should paint a general picture to say why it's important, and then slowly give a sketch of what was done. Then a larger number of people will say, "Yes, Joe has done that," or "Mary has done that; I really see where it is; yes, Mary really gave a good talk; I understand what Mary has done." The tendency is to give a highly restricted, safe talk; this is usually ineffective. Furthermore, many talks are filled with far too much information. So I say this idea of selling is obvious.

Let me summarize. You've got to work on important problems. I deny that it is all luck, but I admit there is a fair element of luck. I subscribe to Pasteur's "Luck favors the prepared mind." I favor heavily what I did. Friday afternoons for years—great thoughts only—means that I committed 10% of my time trying to understand the bigger problems in the field, i.e. what was and what was not important. I found in the early days I had believed 'this' and yet had spent all week marching in 'that' direction. It was kind of foolish. If I really believe the action is over there, why do I march in this direction? I either had to change my goal or change what I did. So I changed something I did and I marched in the direction I thought was important. It's that easy.

Now you might tell me you haven't got control over what you have to work on. Well, when you first begin, you may not. But once you're moderately successful, there are more people asking for results than you can deliver and you have some power of choice, but not completely. I'll tell you a story about that, and it bears on the subject of educating your boss. I had a boss named Schelkunoff; he was, and still is, a very good friend of mine. Some military person came to me and demanded some answers by Friday. Well, I had already dedicated my computing resources to reducing data on the fly for a group of scientists; I was knee deep in short, small, important problems. This military person wanted me to solve his problem by the end of the day on Friday. I said, "No, I'll give it to you Monday. I can work on it over the weekend. I'm not going to do it now." He goes down to my boss, Schelkunoff, and Schelkunoff says, "You must run this for him; he's got to have it by Friday." I tell him, "Why do I? "; he says, "You have to." I said, "Fine, Sergei, but you're sitting in your office Friday afternoon catching the late bus home to watch as this fellow walks out that door." I gave the military person the answers late Friday afternoon. I then went to Schelkunoff's office and sat down; as the man goes out I say, "You see Schelkunoff, this fellow has nothing under his arm; but I gave him the answers." On Monday morning Schelkunoff called him up and said, "Did you come in to work over the weekend?" I could hear, as it were, a pause as the fellow ran through his mind of what was going to happen; but he knew he would have had to sign in, and he'd better not say he had when he hadn't, so he said he hadn't. Ever after that Schelkunoff said, "You set your deadlines; you can change them."

One lesson was sufficient to educate my boss as to why I didn't want to do big jobs that displaced exploratory research and why I was justified in not doing crash jobs which absorb all the research computing facilities. I wanted instead to use the facilities to compute a large number of small problems. Again, in the early days, I was limited in computing capacity and it was clear, in my area, that a "mathematician had no use for machines." But I needed more machine capacity. Every time I had to tell some scientist in some other area, "No I can't; I haven't the machine capacity," he complained. I said "Go tell *your* Vice President that Hamming needs more computing capacity." After a while I could see what was happening up there at the top; many people said to my Vice President, "Your man needs more computing capacity." I got it!

I also did a second thing. When I loaned what little programming power we had to help in the early days of computing, I said,

"We are not getting the recognition for our programmers that they deserve. When you publish a paper you will thank that programmer or you aren't getting any more help from me. That programmer is going to be thanked by name; she's worked hard." I waited a couple of years. I then went through a year of BSTJ articles and counted what fraction thanked some programmer. I took it into the boss and said, "That's the central role computing is playing in Bell Labs; if the BSTJ is important, that's how important computing is." He had to give in. You can educate your bosses. It's a hard job. In this talk I'm only viewing from the bottom up; I'm not viewing from the top down. But I am telling you how you can get what you want in spite of top management. You have to sell your ideas there also.

Well I now come down to the topic, "Is the effort to be a great scientist worth it?" To answer this, you must ask people. When you get beyond their modesty, most people will say, "Yes, doing really first-class work, and knowing it, is as good as wine, women and song put together," or if it's a woman she says, "It is as good as wine, men and song put together." And if you look at the bosses, they tend to come back or ask for reports, trying to participate in those moments of discovery. They're always in the way. So evidently those who have done it, want to do it again. But it is a limited survey. I have never dared to go out and ask those who didn't do great work how they felt about the matter. It's a biased sample, but I still think it is worth the struggle. I think it is very definitely worth the struggle to try and do first-class work because the truth is, the value is in the struggle more than it is in the result. The struggle to make something of yourself seems to be worthwhile in itself. The success and fame are sort of dividends, in my opinion.

I've told you how to do it. It is so easy, so why do so many people, with all their talents, fail? For example, my opinion, to this day, is that there are in the mathematics department at Bell Labs quite a few people far more able and far better endowed than I, but they didn't produce as much. Some of them did produce more than I did; Shannon produced more than I did, and some others produced a lot, but I was highly productive against a lot of other fellows who were better equipped. Why is it so? What happened to them? Why do so many of the people who have great promise, fail?

Well, one of the reasons is drive and commitment. The people who do great work with less ability but who are committed to it, get more done that those who have great skill and dabble in it, who work during the day and go home and do other things and come back and work the next day. They don't have the deep commitment that is apparently necessary for really first-class work. They turn out lots of good work, but we were talking, remember, about first-class work. There is a difference. Good people, very talented people, almost always turn out good work. We're talking about the outstanding work, the type of work that gets the Nobel Prize and gets recognition.

The second thing is, I think, the problem of personality defects. Now I'll cite a fellow whom I met out in Irvine. He had been the head of a computing center and he was temporarily on assignment as a special assistant to the president of the university. It was obvious he had a job with a great future. He took me into his office one time and showed me his method of getting letters done and how he took care of his correspondence. He pointed out how inefficient the secretary was. He kept all his letters stacked around there; he knew where everything was. And he would, on his word processor, get the letter out. He was bragging how marvelous it was and how he could get so much more work done without the secretary's interference. Well, behind his back, I talked to the secretary. The secretary said, "Of course I can't help him; I don't get his mail. He won't give me the stuff to log in; I don't know where he puts it on the floor. Of course I can't help him." So I went to him and said, "Look, if you adopt the present method and do what you can do single-handedly, you can go just that far and no farther than you can do single-handedly. If you will learn to work with the system, you can go as far as the system will support you." And, he never went any further. He had his personality defect of wanting total control and was not willing to recognize that you need the support of the system.

You find this happening again and again; good scientists will fight the system rather than learn to work with the system and take advantage of all the system has to offer. It has a lot, if you learn how to use it. It takes patience, but you can learn how to use the system pretty well, and you can learn how to get around it. After all, if you want a decision 'No', you just go to your boss and get a 'No' easy. If you want to do something, don't ask, do it. Present him with an accomplished fact. Don't give him a chance to tell you 'No'. But if you want a 'No', it's easy to get a 'No'.

Another personality defect is ego assertion and I'll speak in this case of my own experience. I came from Los Alamos and in the early days I was using a machine in New York at 590 Madison Avenue where we merely rented time. I was still dressing in western clothes, big slash pockets, a bolo and all those things. I vaguely noticed that I was not getting as good service as other people. So I set out to measure. You came in and you waited for your turn; I felt I was not getting a fair deal. I said to myself, "Why? No Vice President at IBM said, 'Give Hamming a bad time'. It is the secretaries at the bottom who are doing this. When a slot appears, they'll rush to find someone to slip in, but they go out and find somebody else. Now, why? I haven't mistreated them." Answer, I wasn't dressing the way they felt somebody in that situation should. It came down to just that—I wasn't dressing properly. I had to make the decision—was I going to assert my ego and dress the way I wanted to and have it steadily drain my effort from my professional life, or was I going to appear to conform better? I decided I would make an effort to appear to conform properly. The moment I did, I got much better service. And now, as an old colorful character, I get better service than other people.

You should dress according to the expectations of the audience spoken to. If I am going to give an address at the MIT computer center, I dress with a bolo and an old corduroy jacket or something else. I know enough not to let my clothes, my appearance, my manners get in the way of what I care about. An enormous number of scientists feel they must assert their ego and do their thing their way. They have got to be able to do this, that, or the other thing, and they pay a steady price.

John Tukey almost always dressed very casually. He would go into an important office and it would take a long time before the other fellow realized that this is a first-class man and he had better listen. For a long time John has had to overcome this kind of hostility. It's wasted effort! I didn't say you should conform; I said "The *appearance of conforming* gets you a long way." If you chose to assert your ego in any number of ways, "I am going to do it my way," you pay a small steady price throughout the whole of your professional career. And this, over a whole lifetime, adds up to an enormous amount of needless trouble.

By taking the trouble to tell jokes to the secretaries and being a little friendly, I got superb secretarial help. For instance, one time for some idiot reason all the reproducing services at Murray Hill were tied up. Don't ask me how, but they were. I wanted something done. My secretary called up somebody at Holmdel, hopped the company car, made the hour-long trip down and got it reproduced, and then came back. It was a payoff for the times I had made an effort to cheer her up, tell her jokes and be friendly; it was that little extra work that later paid off for me. By realizing you have to use the system and studying how to get the system to do your work, you learn how to adapt the system to your desires. Or you can fight it steadily, as a small undeclared war, for the whole of your life.

And I think John Tukey paid a terrible price needlessly. He was a genius anyhow, but I think it would have been far better, and far simpler, had he been willing to conform a little bit instead of ego asserting. He is going to dress the way he wants all of the time. It applies not only to dress but to a thousand other things; people will continue to fight the system. Not that you shouldn't occasionally!

When they moved the library from the middle of Murray Hill to the far end, a friend of mine put in a request for a bicycle. Well, the organization was not dumb. They waited awhile and sent back a map of the grounds saying, "Will you please indicate on this map what paths you are going to take so we can get an insurance policy covering you." A few more weeks went by. They then asked, "Where are you going to store the bicycle and how will it be locked so we can do so and so." He finally realized that of course he was going to be red-taped to death so he gave in. He rose to be the President of Bell Laboratories.

Barney Oliver was a good man. He wrote a letter one time to the IEEE. At that time the official shelf space at Bell Labs was so much and the height of the IEEE Proceedings at that time was larger; and since you couldn't change the size of the official shelf space he wrote this letter to the IEEE Publication person saying, "Since so many IEEE members were at Bell Labs and since the official space was so high the journal size should be changed." He sent it for his boss's signature. Back came a carbon with his signature, but he still doesn't know whether the original was sent or not. I am not saying you shouldn't make gestures of reform. I am saying that my study of able people is that they don't get themselves *committed* to that kind of warfare. They play it a little bit and drop it and get on with their work.

Many a second-rate fellow gets caught up in some little twitting of the system, and carries it through to warfare. He expends his energy in a foolish project. Now you are going to tell me that somebody has to change the system. I agree; somebody's has to. Which do you want to be? The person who changes the system or the person who does first-class science? Which person is it that you want to be? Be clear, when you fight the system and struggle with it, what you are doing, how far to go out of amusement, and how much to waste your effort fighting the system. My advice is to let somebody else do it and you get on with becoming a first-class scientist. Very few of you have the ability to both reform the system *and* become a first-class scientist.

On the other hand, we can't always give in. There are times when a certain amount of rebellion is sensible. I have observed almost all scientists enjoy a certain amount of twitting the system for the sheer love of it. What it comes down to basically is that you cannot be original in one area without having originality in others. Originality is being different. You can't be an original scientist without having some other original characteristics. But many a scientist has let his quirks in other places make him pay a far higher price than is necessary for the ego satisfaction he or she gets. I'm not against all ego assertion; I'm against some.

Another fault is anger. Often a scientist becomes angry, and this is no way to handle things. Amusement, yes, anger, no. Anger is misdirected. You should follow and cooperate rather than struggle against the system all the time.

Another thing you should look for is the positive side of things instead of the negative. I have already given you several examples, and there are many, many more; how, given the situation, by changing the way I looked at it, I converted what was apparently a defect to an asset. I'll give you another example. I am an egotistical person; there is no doubt about it. I knew that most people who took a sabbatical to write a book, didn't finish it on time. So before I left, I told all my friends that when I come back, that book was going to be done! Yes, I would have it done—I'd have been ashamed to come back without it! I used my ego to make myself behave the way I wanted to. I bragged about something so I'd have to perform. I found out many times, like a cornered rat in a real trap, I was surprisingly capable. I have found that it paid to say, "Oh yes, I'll get the answer for you Tuesday," not having any idea how to do it. By Sunday night I was really hard thinking on how I was going to deliver by Tuesday. I often put my pride on the line and sometimes I failed, but as I said, like a cornered rat I'm surprised how often I did a good job. I think you need to learn to use yourself. I think you need to know how to convert a situation from one view to another which would increase the chance of success.

Now self-delusion in humans is very, very common. There are enumerable ways of you changing a thing and kidding yourself and making it look some other way. When you ask, "Why didn't you do such and such," the person has a thousand alibis. If you look at the history of science, usually these days there are 10 people right there ready, and we pay off for the person who is there first. The other nine fellows say, "Well, I had the idea but I didn't do it and so on and so on." There are so many alibis. Why weren't you first? Why didn't you do it right? Don't try an alibi. Don't try and kid yourself. You can tell other people all the alibis you want. I don't mind. But to yourself try to be honest.

If you really want to be a first-class scientist you need to know yourself, your weaknesses, your strengths, and your bad faults, like my egotism. How can you convert a fault to an asset? How can you convert a situation where you haven't got enough manpower to move into a direction when that's exactly what you need to do? I say again that I have seen, as I studied the history, the successful scientist changed the viewpoint and what was a defect became an asset.

In summary, I claim that some of the reasons why so many people who have greatness within their grasp don't succeed are: they don't work on important problems, they don't become emotionally involved, they don't try and change what is difficult to some other situation which is easily done but is still important, and they keep giving themselves alibis why they don't. They keep saying that it is a matter of luck. I've told you how easy it is; furthermore I've told you how to reform. Therefore, go forth and become great scientists!

(End of the formal part of the talk.)

Discussion – Questions and Answers

A.G. Chynoweth: Well that was 50 minutes of concentrated wisdom and observations accumulated over a fantastic career; I lost track of all the observations that were striking home. Some of them are very very timely. One was the plea for more computer capacity; I was hearing nothing but that this morning from several people, over and over again. So that was right on the mark today even though here we are 20 - 30 years after when you were making similar remarks, Dick. I can think of all sorts of lessons that all of us can draw from your talk. And for one, as I walk around the halls in the future I hope I won't see as many closed doors in Bellcore. That was one observation I thought was very intriguing. Thank you very, very much indeed Dick; that was a wonderful recollection. I'll now open it up for questions. I'm sure there are many people who would like to take up on some of the points that Dick was making.

Hamming: First let me respond to Alan Chynoweth about computing. I had computing in research and for 10 years I kept telling my management, "Get that !&@#% machine out of research. We are being forced to run problems all the time. We can't do research because were too busy operating and running the computing machines." Finally the message got through. They were going to move computing out of research to someplace else. I was persona non grata to say the least and I was surprised that people didn't kick my shins because everybody was having their toy taken away from them. I went in to Ed David's office and said, "Look Ed, you've got to give your researchers a machine. If you give them a great big machine, we'll be back in the same trouble we were before, so busy keeping it going we can't think. Give them the smallest machine you can because they are very able people. They will learn how to do things on a small machine instead of mass computing." As far as I'm concerned, that's how UNIX arose. We gave them a moderately small machine and they decided to make it do great things. They had to come up with a system to do it on. It is called UNIX!

A.G. Chynoweth: I just have to pick up on that one. In our present environment, Dick, while we wrestle with some of the red tape attributed to, or required by, the regulators, there is one quote that one exasperated AVP came up with and I've used it over and over again. He growled that, "UNIX was never a deliverable! "

Question: What about personal stress? Does that seem to make a difference?

Hamming: Yes, it does. If you don't get emotionally involved, it doesn't. I had incipient ulcers most of the years that I was at Bell Labs. I have since gone off to the Naval Postgraduate School and laid back somewhat, and now my health is much better. But if you want to be a great scientist you're going to have to put up with stress. You can lead a nice life; you can be a nice guy or you can be a great scientist. But nice guys end last, is what Leo Durocher said. If you want to lead a nice happy life with a lot of recreation and everything else, you'll lead a nice life.

Question: The remarks about having courage, no one could argue with; but those of us who have gray hairs or who are well established don't have to worry too much. But what I sense among the young people these days is a real concern over the risk taking in a highly competitive environment. Do you have any words of wisdom on this?

Hamming: I'll quote Ed David more. Ed David was concerned about the general loss of nerve in our society. It does seem to me that we've gone through various periods. Coming out of the war, coming out of Los Alamos where we built the bomb, coming out of building the radars and so on, there came into the mathematics department, and the research area, a group of people with a lot of guts. They've just seen things done; they've just won a war which was fantastic. We had reasons for having courage and therefore we did a great deal. I can't arrange that situation to do it again. I cannot blame the present generation for not having it, but I agree with what you say; I just cannot attach blame to it. It doesn't seem to me they have the desire for greatness; they lack the courage to do it. But we had, because we were in a favorable circumstance to have it; we just came through a tremendously successful war. In the war we were looking very, very bad for a long while; it was a very desperate struggle as you well know. And our success, I think, gave us courage and self confidence; that's why you see, beginning in the late forties through the fifties, a tremendous productivity at the labs which was stimulated from the earlier times. Because many of us were earlier forced to learn other things—we were forced to learn the things we didn't want to learn, we were forced to have an open door—and then we could exploit those things we learned. It is true, and I can't do anything about it; I cannot blame the present generation either. It's just a fact.

Question: Is there something management could or should do?

Hamming: Management can do very little. If you want to talk about managing research, that's a totally different talk. I'd take another hour doing that. This talk is about how the individual gets very successful research done in spite of anything the management does or in spite of any other opposition. And how do you do it? Just as I observe people doing it. It's just that simple and that hard!

Question: Is brainstorming a daily process?

Hamming: Once that was a very popular thing, but it seems not to have paid off. For myself I find it desirable to talk to other people; but a session of brainstorming is seldom worthwhile. I do go in to strictly talk to somebody and say, "Look, I think there has to be something here. Here's what I think I see…" and then begin talking back and forth. But you want to pick capable people. To use another analogy, you know the idea called the 'critical mass.' If you have enough stuff you have critical mass. There is also the idea I used to call 'sound absorbers'. When you get too many sound absorbers, you give out an idea and they merely say, "Yes, yes, yes." What you want to do is get that critical mass in action; "Yes, that reminds me of so and so," or, "Have you thought about that or this?" When you talk to other people, you want to get rid of those sound absorbers who are nice people but merely say, "Oh yes," and to find those who will stimulate you right back.

For example, you couldn't talk to John Pierce without being stimulated very quickly. There were a group of other people I used to talk with. For example there was Ed Gilbert; I used to go down to his office regularly and ask him questions and listen and come back stimulated. I picked my people carefully with whom I did or whom I didn't brainstorm because the sound absorbers are a curse. They are just nice guys; they fill the whole space and they contribute nothing except they absorb ideas and the new ideas just die away instead of echoing on. Yes, I find it necessary to talk to people. I think people with closed doors fail to do this so they fail to get their ideas sharpened, such as "Did you ever notice something over here?" I never knew anything about it—I can go over and look. Somebody points the way. On my visit here, I have already found several books that I must read when I get home. I talk to people and ask questions when I think they can answer me and give me clues that I do not know about. I go out and look!

Question: What kind of tradeoffs did you make in allocating your time for reading and writing and actually doing research?

Hamming: I believed, in my early days, that you should spend at least as much time in the polish and presentation as you did in the original research. Now at least 50% of the time must go for the presentation. It's a big, big number.

Question: How much effort should go into library work?

Hamming: It depends upon the field. I will say this about it. There was a fellow at Bell Labs, a very, very, smart guy. He was always in the library; he read everything. If you wanted references, you went to him and he gave you all kinds of references. But in the middle of forming these theories, I formed a proposition: there would be no effect named after him in the long run. He is now retired from Bell Labs and is an Adjunct Professor. He was very valuable; I'm not questioning that. He wrote some very good Physical Review articles; but there's no effect named after him because he read too much. If you read all the time what other people have done you will think the way they thought. If you want to think new thoughts that are different, then do what a lot of creative people do—get the problem reasonably clear and then refuse to look at any answers until you've thought the problem through carefully how you would do it, how you could slightly change the problem to be the correct one. So yes, you need to keep up. You need to keep up more to find out what the problems are than to read to find the solutions. The reading is necessary to know what is going on and what is possible. But reading to get the solutions does not seem to be the way to do great research. So I'll give you two answers. You read; but it is not the amount, it is the way you read that counts.

Question: How do you get your name attached to things?

Hamming: By doing great work. I'll tell you the hamming window one. I had given Tukey a hard time, quite a few times, and I got a phone call from him from Princeton to me at Murray Hill. I knew that he was writing up power spectra and he asked me if I would mind if he called a certain window a "Hamming window." And I said to him, "Come on, John; you know perfectly well I did only a small part of the work but you also did a lot." He said, "Yes, Hamming, but you contributed a lot of small things; you're entitled to some credit." So he called it the hamming window. Now, let me go on. I had twitted John frequently about true greatness. I said true greatness is when your name is like ampere, watt, and fourier—when it's spelled with a lower case letter. That's how the hamming window came about.

Question: Dick, would you care to comment on the relative effectiveness between giving talks, writing papers, and writing books?

Hamming: In the short-haul, papers are very important if you want to stimulate someone tomorrow. If you want to get recognition long-haul, it seems to me writing books is more contribution because most of us need orientation. In this day of practically infinite knowledge, we need orientation to find our way. Let me tell you what infinite knowledge is. Since from the time of Newton to now, we have come close to doubling knowledge every 17 years, more or less. And we cope with that, essentially, by specialization. In the next 340 years at that rate, there will be 20 doublings, i.e. a million, and there will be a million fields of specialty for every one field now. It isn't going to happen. The present growth of knowledge will choke itself off until we get different tools. I believe that books which try to digest, coordinate, get rid of the duplication, get rid of the less fruitful methods and present the underlying ideas clearly of what we know now, will be the things the future generations will value. Public talks are necessary; private talks are necessary; written papers are necessary. But I am inclined to believe that, in the long-haul, books which leave out what's not essential are more important than books which tell you everything because you don't want to know everything. I don't want to know that much about penguins is the usual reply. You just want to know the essence.

Question: You mentioned the problem of the Nobel Prize and the subsequent notoriety of what was done to some of the careers. Isn't that kind of a much more broad problem of fame? What can one do?

Hamming: Some things you could do are the following. Somewhere around every seven years make a significant, if not complete, shift in your field. Thus, I shifted from numerical analysis, to hardware, to software, and so on, periodically, because you tend to use up your ideas. When you go to a new field, you have to start over as a baby. You are no longer the big mukity muk and you can start back there and you can start planting those acorns which will become the giant oaks. Shannon, I believe, ruined himself. In fact when he left Bell Labs, I said, "That's the end of Shannon's scientific career." I received a lot of flak from my friends who said that Shannon was just as smart as ever. I said, "Yes, he'll be just as smart, but that's the end of his scientific career," and I truly believe it was.

You have to change. You get tired after a while; you use up your originality in one field. You need to get something nearby. I'm not saying that you shift from music to theoretical physics to English literature; I mean within your field you should shift areas so that you don't go stale. You couldn't get away with forcing a change every seven years, but if you could, I would require a condition for doing research, being that you will change your field of research every seven years with a reasonable definition of what it means, or at the end of 10 years, management has the right to compel you to change. I would insist on a change because I'm serious. What happens to the old fellows is that they get a technique going; they keep on using it. They were marching in that direction which was right then, but the world changes. There's the new direction; but the old fellows are still marching in their former direction.

You need to get into a new field to get new viewpoints, and *before* you use up all the old ones. You can do something about this, but it takes effort and energy. It takes courage to say, "Yes, I will give up my great reputation." For example, when error correcting codes were well launched, having these theories, I said, "Hamming, you are going to quit reading papers in the field; you are going to ignore it completely; you are going to try and do something else other than coast on that." I deliberately refused to go on in that field. I wouldn't even read papers to try to force myself to have a chance to do something else. I managed myself, which is what I'm preaching in this whole talk. Knowing many of my own faults, I manage myself. I have a lot of faults, so I've got a lot of problems, i.e. a lot of possibilities of management.

Question: Would you compare research and management?

Hamming: If you want to be a great researcher, you won't make it being president of the company. If you want to be president of the company, that's another thing. I'm not against being president of the company. I just don't want to be. I think Ian Ross does a good job as President of Bell Labs. I'm not against it; but you have to be clear on what you want. Furthermore, when you're young, you may have picked wanting to be a great scientist, but as you live longer, you may change your mind. For instance, I went to my boss, Bode, one day and said, "Why did you ever become department head? Why didn't you just be a good scientist?" He said, "Hamming, I had a vision of what mathematics should be in Bell Laboratories. And I saw if that vision was going to be realized, *I* had to make it happen; *I* had to be department head." When your vision of what you want to do is what you can do single-handedly, then you should pursue it. The day your vision, what you think needs to be done, is bigger than what you can do single-handedly, then you have to move toward management. And the bigger the vision is, the farther in management you have to go. If you have a vision of what the whole laboratory should be, or the whole Bell System, you have to get there to make it happen. You can't make it happen from the bottom very easily. It depends upon what goals and what desires you have. And as they change in life, you have to be prepared to change. I chose to avoid management because I preferred to do what I could do single-handedly. But that's the choice that I made, and it is biased. Each person is entitled to their choice. Keep an open mind. But when you do choose a path, for heaven's sake be aware of what you have done and the choice you have made. Don't try to do both sides.

Question: How important is one's own expectation or how important is it to be in a group or surrounded by people who expect great work from you?

Hamming: At Bell Labs everyone expected good work from me—it was a big help. Everybody expects you to do a good job, so you do, if you've got pride. I think it's very valuable to have first-class people around. I sought out the best people. The moment that physics table lost the best people, I left. The moment I saw that the same was true of the chemistry table, I left. I tried to go with people who had great ability so I could learn from them and who would expect great results out of me. By deliberately managing myself, I think I did much better than laissez faire.

Question: You, at the outset of your talk, minimized or played down luck; but you seemed also to gloss over the circumstances that got you to Los Alamos, that got you to Chicago, that got you to Bell Laboratories.

Hamming: There was some luck. On the other hand I don't know the alternate branches. Until you can say that the other branches would not have been equally or more successful, I can't say. Is it luck the particular thing you do? For example, when I met Feynman at Los Alamos, I knew he was going to get a Nobel Prize. I didn't know what for. But I knew darn well he was going to do great work. No matter what directions came up in the future, this man would do great work. And sure enough, he did do great work. It isn't that you only do a little great work at this circumstance and that was luck, there are many opportunities sooner or later. There are a whole pail full of opportunities, of which, if you're in this situation, you seize one and you're great over there instead of over here. There is an element of luck, yes and no. Luck favors a prepared mind; luck favors a prepared person. It is not guaranteed; I don't guarantee success as being absolutely certain.

I'd say luck changes the odds, but there is some definite control on the part of the individual.

Go forth, then, and do great work!

(End of the General Research Colloquium Talk.)

Acknowledgment

I would like to acknowledge the professional efforts of Donna Paradise of the Word Processing Center who did the initial transcription of the talk from the tape recording. She made my job of editing much easier. The errors of sentence parsing and punctuation are mine and mine alone. Finally I would like to express my sincere appreciation to Richard Hamming and Alan Chynoweth for all of their help in bringing this transcription to its present readable state. J.F. Kaiser

17

Ten Simple Rules for Doing Your Best Research, According to Hamming

Thomas C. Erren, Paul Cullen, Michael Erren, Philip E. Bourne

Thomas C. Erren is a physician epidemiologist at the University of Cologne, Köln, Lindenthal, Germany where he is a frequent writer on scientific research topics. Paul Cullen and Michael Erren are with the Institute of Clinical Chemistry and Laboratory Medicine, University of Münster, Münster, Germany where they inter alia research the vascular system.

This editorial can be considered the preface to the "Ten Simple Rules" series [1–7]. The rules presented here are somewhat philosophical and behavioural rather than concrete suggestions for how to tackle a particular scientific professional activity such as writing a paper or a grant. The thoughts presented are not our own; rather, we condense and annotate some excellent and timeless suggestions made by the mathematician Richard Hamming two decades ago on how to do "first-class research." [8] As far as we know, the transcript of the Bell Communications Research Colloquium Seminar provided by Dr. Kaiser [8] was never formally published, so that Dr. Hamming's thoughts are not as widely known as they deserve to be. By distilling these thoughts into something that can be thought of as "Ten Simple Rules," we hope to bring these ideas to broader attention.

Hamming's 1986 talk was remarkable. In "You and Your Research," he addressed the question: How can scientists do great re-

* Reprinted from Erren TC, Cullen P, Erren M, Bourne PE,"Ten Simple Rules for Doing Your Best Research, According to Hamming" *PLoS Computational Biology* Vol. 3 No. 10: e213 (2007); Creative Commons Attribution License

search, i.e., Nobel-Prize-type work? His insights were based on more than forty years of research as a pioneer of computer science and telecommunications who had the privilege of interacting with such luminaries as the physicists Richard Feynman, Enrico Fermi, Edward Teller, Robert Oppenheimer, Hans Bethe, and Walter Brattain, with Claude Shannon, "the father of information theory," and with the statistician John Tukey. Hamming "became very interested in the difference between those who do and those who might have done," and he offered a number of answers to the question "why…so few scientists make significant contributions and so many are forgotten in the long run?" We have condensed Hamming's talk into the ten rules listed below:

Rule 1: Drop Modesty

To quote Hamming: "Say to yourself: 'Yes, I would like to do first-class work.' Our society frowns on people who set out to do really good work. But you should say to yourself: 'Yes, I would like to do something significant.'"

Rule 2: Prepare Your Mind

Many think that great science is the result of good luck, but luck is nothing but the marriage of opportunity and preparation. Hamming cites Pasteur's adage that "luck favours the prepared mind."

Rule 3: Age Is Important

Einstein did things very early, and all the "quantum mechanics fellows," as well as most mathematicians and astrophysicists, were, as Hamming notes, "disgustingly young" when they did their best work. On the other hand, in the fields of music, politics, and literature, the protagonists often produce what we consider their best work late in life.

Rule 4: Brains Are Not Enough, You Also Need Courage

Great scientists have more than just brainpower. To again cite Hamming: "Once you get your courage up and believe that you can do important things, then you can. If you think you can't, almost surely you are not going to. Great scientists will go forward under incredible circumstances; they think and continue to think."

Rule 5: Make the Best of Your Working Conditions

To paraphrase Hamming, what most people think are the best working conditions clearly are not, because people are often most productive when working conditions are bad. One of the better times of the Cambridge Physical Laboratories was when they worked practically in shacks—they did some of the best physics ever. By turning the problem around a bit, great scientists often transform an apparent defect into an asset. "It is a poor workman who blames his tools—the good man gets on with the job, given what he's got, and gets the best answer he can."

Rule 6: Work Hard and Effectively

Most great scientists have tremendous drive, and most of us would be surprised how much we would know if we worked as hard as some great scientists did for many years. As Hamming says: "Knowledge and productivity are like compound interest. Given two people with exactly the same ability, the one person who manages day in and day out to get in one more hour of thinking will be tremendously more productive over a lifetime." But, Hamming notes, hard work alone is not enough—it must be applied sensibly.

Rule 7: Believe and Doubt Your Hypothesis at the Same Time

Great scientists tolerate ambiguity. They believe the theory enough to go ahead; they doubt it enough to notice the errors and faults so they can step forward and create the new replacement theory. As Hamming says: "When you find apparent flaws, you've got to be sensitive and keep track of those things, and keep an eye out for how they can be explained or how the theory can be changed to fit them. Those are often the great scientific contributions."

Rule 8: Work on the Important Problems in Your Field

It is surprising but true that the average scientist spends almost all his time working on problems that he believes not to be important and not to be likely to lead to important results. By contrast, those seeking to do great work must ask: "What are the important problems of my field? What important problems am I working on?" Hamming again: "It's that simple. If you want to do great work, you clearly must work on important problems…I finally adopted what I called 'Great Thoughts Time.' When I went to lunch Friday noon, I would only discuss great thoughts after that. By great thoughts I mean ones like: 'What will be the impact of computers on science and how can I change it?'"

Rule 9: Be Committed to Your Problem

Scientists who are not fully committed to their problem seldom produce first-class work. To a large extent, creativity comes out of the subconscious. If you are deeply immersed in and committed to a topic, day after day, your subconscious has nothing to do but work on your problem. Hamming says it best: "So the way to manage yourself is that when you have a real important problem you don't let anything else get the center of your attention—you keep your thoughts on the problem. Keep your subconscious starved so it has to work on your problem, so you can sleep peacefully and get the answer in the morning, free."

Rule 10: Leave Your Door Open

Keeping the door to your office closed makes you more productive in the short term. But ten years later, somehow you may not quite know what problems are worth working on, and all the hard work you do will be "sort of tangential" in importance. He (or she) who leaves the door open gets all kinds of interruptions, but he (or she) also occasionally gets clues as to what the world is and what might be important. Again, Hamming deserves to be quoted verbatim: "There is a pretty good correlation between those who work with the doors open and those who ultimately do important things, although people who work with doors closed often work harder. Somehow they seem to work on slightly the wrong thing—not much, but enough that they miss fame."

In our view, Rule 10 may be the key to getting the best research done because it will help you to obey Rules 1–9, and, most importantly, it will foster group creativity [9]. A discussion over lunch with your colleagues is often worth much more than a trip to the library. However, when choosing your lunchmates (and, by implication, your institution), be on your toes. As Hamming says: "When you talk to other people, you want to get rid of those sound absorbers who are nice people but merely say 'Oh yes,' and to find those who will stimulate you right back."

References

[1] Bourne, P.E., "Ten simple rules for getting published." *PLoS Computational Biology,* Vol. 1 e57 (2005).

[2] Bourne, P.E. & Chalupa, L.M., "Ten simple rules for getting grants." *PLoS Computational Biology,* Vol. 2 e12 (2006).

[3] Bourne, P.E. & Korngreen, A. "Ten simple rules for reviewers." *PLoS Computational Biology,* Vol. 2, e110 (2006).

[4] Bourne, P.E. & Friedberg, I., "Ten simple rules for selecting a postdoctoral position." *PLoS Computational Biology,* Vol. 2, e121 (2006).

[5] Vicens, Q. & Bourne, P.E., "Ten simple rules for a successful collaboration." *PLoS Computational Biology,* Vol. 3, e44 (2007).

[6] Bourne, P.E., "Ten simple rules for making good oral presentations." *PLoS Computational Biology,* Vol. 3, e77 (2007).

[7] Erren, T.C. & Bourne, P.E., "Ten simple rules for a good poster presentation." *PLoS Computational Biology,* Vol. 3, e102 (2007).

[8] Hamming, R., "You and your research." In: Ed. Kaiser, J.F., Transcription of the Bell Communications Research Colloquium Seminar; 7 March 1986; Morristown, New Jersey, United States.

[9] Erren, T.C., "Hamming's 'open doors' and group creativity as keys to scientific excellence: The example of Cambridge." *Medical Hypotheses,* Vol. 3, 17804173 (2007).

18

Memoirs of a Dissident Scientist Hannes Alfvén

Hannes Alfvén (May 30, 1908 – April 2, 1995)—was a Swedish electrical engineer turned plasma physicist who is considered one of the giants of this field having predicted many phenomena such as magnetohydrodynamic waves and the interstellar magnetic field. He was awarded the Nobel Prize in 1970 for his work on plasma physics. He also held several unorthodox (and since disproven) scientific opinions such as matter-antimatter symmetry in the universe and a local origin for cosmic rays. In this article he describes his views on creativity and scientific independence.

Half a century ago I started like many other young physicists in nuclear physics. As I was more interested in experiments than theory, my first work in this field was concentrated on electrical methods to detect nuclear radiations. That led me to work on Geiger-Müller tubes and I constructed giant tubes which were suited for measuring cosmic rays. I found this field of investigation to be a fascinating field because cosmic rays had energies far larger than the radiations from radioactive substances, indeed up to 10^{10} or perhaps even 10^{11} eV! I began to speculate about their origin and made an ingenious theory which in reality was so silly that I do not even mention it here. For unknown reasons *Nature* agreed to publish a letter about it, but as soon as it was published I found that the idea was completely unreasonable.

This must have been around 1933, because I remember very well that it was before the first international conference I ever attended which was in London 1934. At that conference I met A.H. Compton, who immediately mentioned that he had read my letter and thought it interesting. I said that I was ashamed of having

* Reprinted with permission of the publisher from Hannes Alfvén,"Memoirs of a Dissident Scientist" In: Ed. Y. Sekido and H. Elliot *Early History of Cosmic Ray Studies* (Dordrecht: Reidel, 1985) 427-431.

published something so completely silly to which he answered: "Don't give it up too easily." As he was one of the great authorities on cosmic radiation this was an enormous encouragement to me. In retrospect this may have been the trigger which brought me into astrophysics. It happened at a time when almost everybody was running toward nuclear physics. It saved me from the guilt associated with atomic bombs and nuclear energy which every nuclear physicist of today must feel at the bottom of his heart.

Next episode is dated 1948. In the meantime, there had been a war, the atomic bombs had been exploded and hence, science had changed its character forever. My scientific work in astrophysics had continued. I had presented a theory of cosmic ray acceleration by electromagnetic effects around double stars. With the exception of Swann's famous but unrealistic "cygnotron" (as Vallarta called it) I think this was the first attempt to explain the high energies as due to electromagnetic effects. However, processes of this kind could not supply enough intensity if cosmic radiation filled the whole universe, as was the generally accepted view. I then pointed out that this dilemma could be solved if cosmic radiation was confined to our galaxy. This required a galactic magnetic field of at least 10^{-10} gauss. (It should be remembered that the highest cosmic radiation energies known at that time were only 10^{11} eV.)

I tried to publish this paper in—I think—Nature and other journals which at that time were generally read, but it was not accepted. "Is that guy completely crazy? Doesn't he know that space is empty which means that there can be no particles carrying currents. And we all know that the earth's magnetic field derives from a permanent magnet in its center. Hasn't he understood how rapidly it decreases with distance? Does he believe that there are a number of magnets floating around in the galaxy?" The only journal which accepted it was *Arkiv for Fysik*, to which it was communicated by my teacher Professor Manne Siegbahn. Certainly this was a very reputable journal but not very much read. So with this as a background, I had the following experience in 1948. I was starting my first transatlantic journey and on my way I attended a conference in Birmingham. I came a little late to the meeting and entered a lecture hall where Edward Teller spoke about the origin of cosmic rays. I had shown beyond any doubt—I thought—hat cosmic radiation was a galactic phenomenon generated by some electromagnetic effects and trapped inside the galaxy where its intensity of course was uniform. Teller claimed that it was a local phenomenon, generated and trapped inside what we today call the heliosphere. The whole audience roared with laughter, and so did I. It was the first time I heard Teller and did not know that this dynamic personality always makes everybody laugh—independent of whether he speaks about his dear atomic bombs or astrophysics.

When I came to the USA, quite a few people asked me about Teller's new idea. I said that everybody laughed at his lecture and that this was nothing to be taken seriously. I made the grand tour of the USA—of course quite an experience to anybody who sees it for the first time. I spent some time at Caltech and then Teller came there. To my great surprise he repeated the same silly lecture. Now I objected to it and after his lecture we had a long discussion with the result that he invited me to continue the discussion at his seminar in Chicago on my way back. I must admit that he defended his views in a clever way, and I must admit that it was not at all so obvious that cosmic radiation was a galactic phenomenon as I had thought. After a few days of thinking I actually got convinced that he was essentially correct. I sent him a picture postcard where I said that I agreed in essential respects, and even had found some new arguments in favor of the heliospheric confinement of cosmic rays. I think that these were connected with the magnetohydrodynamic waves which I had published my first paper about in 1942.

When a few weeks later I came to Chicago Teller introduced me at this seminar with the words: "I need only tell you that this is the guy who wrote the picture postcard." Teller and I later wrote a joint paper about the local origin of cosmic rays. After some years Teller changed his views—and has since then believed in the standard views of cosmic radiation as a galactic phenomenon. This is now so generally accepted that the radiation is referred to as the "galactic radiation." It is a little paradoxical that I now should be one of the rather few supporters of Teller's theory, whereas the galactic theory which essentially derives from my views in the late 1930s is generally accepted. In fact it has become so sacrosanct that all my attempts for decades to start a serious discussion have led to nothing.

My brief visit to Chicago in 1948 had also another consequence. One of the members of the seminar was Enrico Fermi, who got interested in the origin of cosmic rays. After the seminar he asked me to explain what the magnetohydrodynamic waves were. Since I published my first paper in 1942 very few people—with Lyman Spitzer and Martin Schwarzschild as the most prominent exceptions—had believed in them. I got letters from colleagues who asked me whether I had not understood that this was nonsense. If they existed, Maxwell would have described them and it is quite clear that he has not. Hence it is impossible that they could exist! I completely agree that they are nothing remarkable. (I considered them mainly as a spin-off of my theory of sunspots, which I thought to be much more important). To discover them would have been an appropriate pastime for Maxwell on a tedious Sunday afternoon, but as a matter of fact he seems to have preferred other pastimes.

Fermi listened to what I said about them for five or ten minutes, and then he said: "Of course such waves could exist." Fermi had such an authority that if he said "of course" today, every physicist said "of course" tomorrow. Actually he published a paper in which he explained them in such a clear way that no one could doubt their possible existence. What I had not succeeded to do in six years was done by Fermi in what was only the introduction to a presentation of the famous Fermi mechanism of cosmic ray acceleration (the importance of which I have not yet been able to understand).

From my personal point of view my early work on cosmic rays was very important. Contrary to almost all astrophysicists my education had taken place in a laboratory where my hands got dirty when I studied electrical discharges. Hence it was natural to me to describe plasma phenomena in terms of particles moving in magnetic fields in the same way they do in the laboratory. The cosmic ray phenomena in space led me to approach the whole of astrophysics in this way. Since then the astrophysics of low density plasmas has been to me a science in which all phenomena ultimately should be described from the point of view of the individual particles. Instead of treating hydromagnetic equations I prefer to sit and ride on each electron and ion and try to imagine what the world is like from its point of view and what forces push them to the left or to the right.

This has been a great advantage because it gives me a possibility to approach the phenomena from another point than most astrophysicists do, and it is always fruitful to look at *any* phenomenon under two different points of view. On the other hand it has given me a serious disadvantage. When I describe the phenomena according to this formalism most referees do not understand what I say and turn down my papers. With the referee system which rules US science today, this means that my papers rarely are accepted by the leading US journals. Europe, including the Soviet Union, and Japan are more tolerant of dissidents.

The hydromagnetic approach to cosmic plasma is probably rather good for high density plasmas like plasmas in stellar interiors, but if applied to low density plasmas in the cosmos it is often terribly misleading. From the thermonuclear crisis 20 years ago and the similar crisis in magnetospheric physics the last five years, we should learn that it is unrealistic to approach a low density plasma by considering it as a fluid. Indeed, the plasma has taught me that it is so complicated that it should be regarded more as a living being than a dead mechanical system. (Langmuir, who baptized it, under stood this and borrowed its name from the blood plasma). This creature does not understand differential equations and vectors and tensors and does not care for such nonsense. It always finds new ways to cheat the mathematical physicist. This means that unless a scientist purges his brain of all such stuff he has little chance of understanding this naughty and whimsical child who loves to revolt against what the theoreticians have pre scribed that it should do.

The mentioned conditions and quite a few other factors have led to a disagreement between a very strong establishment (E) and a small group of dissidents (D) to which the present author belongs. This is nothing remarkable. What is more remarkable and regrettable is that it seems to be almost impossible to start a serious discussion between E and D. As a dissident is in a very unpleasant situation, I am sure that D would be very glad to change their views as soon as E gives convincing arguments. But the argument "all knowledgeable people agree that…" (with the tacit addition that by not agreeing you demonstrate that you are a crank) is not a valid argument in science. If scientific issues always were decided by Gallup polls and not by scientific arguments science will very soon be petrified forever.

At a symposium on radio astronomy in 1966 the following exchange of remarks took place.
H. Alfvén. [V.L.] Ginzburg has said that it is absolutely clear that cosmic radiation plays a decisive role in the galaxy. I am not at all sure about this, because what we observe and what we conclude from observation are so different…It may very well be that 99% of the cosmic radiation is a local phenomenon confined to our environment in the same way as the Van Allen radiation belts are confined to the earth's magnetic field.
Ginzburg answers: The arguments against the solar or local origin of cosmic rays are numerous…The radio astronomical evidence is quite strong…However, I agree that it is extraordinarily difficult to disprove anything.
Alfvén: To disprove anything is very difficult, but also to prove it.
Ginzburg: Fortunately it is possible to do something. I have worked in the field for some years, and I can say in the course of time the argument slowly improves. So I hope during my lifetime I shall see the full victory of these things.
Alfvén: I hope you will live very long.

19

Excerpt from "A Backward Look to the Future" Edwin Thompson Jaynes

Edwin T. Jaynes (July 5, 1922 – April 30, 1998)—was a theoretical physicist at the Washington University in St. Louis. He is well-known for his brilliant expositions on probability and statistical mechanics, showing that equilibrium statistical mechanics could be interpreted as an outgrowth of information theory under the principle of maximum entropy. Maximum entropy and methods of Bayesian inference in the sciences widely owe their debt to him. Fearlessly innovative, he often took iconoclastic stances such as suggesting even thermodynamic manifestations of entropy are due to the ignorance of the observer in their knowledge of the system dynamics when observing macroscopic properties. In this excerpt from one of his essays later in life, he gives advice to starting scientists on how to approach research and weather criticism.

Dealing with Critics

Looking back over the past forty years, I can see that the greatest mistake I made was to listen to the advice of people who were opposed to my efforts. Just at the peak of my powers I lost several irreplaceable years because I allowed myself to become discouraged by the constant stream of criticism from the Establishment, that descended upon everything I did. I have never, except in the past few years, had the slightest encouragement from others to pursue my work; the drive to do it had to come entirely from within me. The result was that my contributions to probability theory were delayed by about a decade, and my potential con-

* Reprinted with permission from Edwin Jaynes, "A Backward Look to the Future" In: Eds. W.T. Grandy, Jr. and P.W. Milonni, *Physics & Probability: Essays in honor of Edwin T. Jaynes* (New York: Cambridge University Press, 1993) 272-273.

tributions to electrodynamics, whatever they might have been, are probably lost forever.

But I can now see that all of this criticism was based on misunderstanding or ideology. My perceived sin was not in my logic or mathematics; it was that I did not subscribe to the dogmas emanating from Copenhagen and Rothamsted. Yet I submit that breaking those dogmas was the necessary prerequisite to making any further progress in quantum theory and probability theory. If not in my way, then necessarily in some other.

In any field, the Establishment is not seeking the truth, because it is composed of those who, having found part of it yesterday, believe that they are in possession of all of it today. Progress requires the introduction, not just of new mathematics which is always tolerated by the Establishment; but new conceptual ideas which are *necessarily different from those* held by the Establishment (for, if the ideas of the Establishment were sufficient to lead to further progress, that progress would have been made).

Therefore, to anyone who has new ideas of a currently unconventional kind, I want to give this advice, in the strongest possible terms: *Do not allow yourself to be discouraged or dejected from your course by negative criticisms*, particularly those that were invented for the sole purpose of discouraging you, unless they exhibit some clear and specific error of reasoning or conflict with experiment. Unless they can do this, your critics are almost certainly wrong, but to reply by trying to show exactly where and why they are wrong would be wasted effort which would not convince your critics and would only keep you from the far more important, constructive things that you might have accomplished in the same time. Let others deal with them; if you allow your enemies to direct your work, then they have won after all.

Although the arguments of your critics are almost certainly wrong, they will retain just enough plausibility in the minds of some to maintain a place for them in the realm of controversy; that is just a fact of life that you must accept as the price of doing creative work. Take comfort in the historical record, which shows that no creative person has ever been able to escape this; the more fundamental the new idea, the more bitter the controversy it will stir up. Newton, Darwin, Boltzmann, Pasteur, Einstein, Wegener were all embroiled in this. Newton wrote in 1676:

> *I see a man must either resolve to put out nothing new, or become a slave to defend it.*

Throughout his lifetime, Alfred Wegener received nothing but attacks on his ideas; yet he was right and today those ideas are the foundation of geophysics. We revere the names of James Clerk Maxwell and J. Willard Gibbs; yet their work was never fully appreciated in their lifetimes, and even today it is still, like that of Darwin, under attack by persons who, after a Century, have not yet comprehended their message [1].

The recent reminiscences of Francis Crick [2] make many other important points that we might otherwise have included here, because they apply as well to physics as to biology. For example, he notes that in studying brain function, theoretical work tended to fall into a number of somewhat separate schools, each of which was rather reluctant to quote the work of the others. This is usually characteristic of a subject that is not producing any definite conclusions." Exactly the same could be said of several areas of physics; we would add only that, even when definite conclusions are proclaimed by one school, they are not often justified or permanent.

References

[1] Atkins, P.W., "Entropy in Relation to Complete Knowledge," *Contemporary Physics.* Vol. 27, 257 (1986).

[2] Crick, F., *What Mad Pursuit*, (New York: Basic Books, Inc., 1988).

20

Autobiographical Sketch Hermann von Helmholtz

Hermann von Helmholtz (August 31, 1821 – September 8, 1894)—was a German scientist who contributed both to physiology and physics. He is best known for his work in thermodynamics including the First Law of Thermodynamics that states the principle of the conservation of energy. He also did work in the physiology of nerves and estimated the speed of light by measuring the speed of electric impulses through a Leyden jar (capacitor). In this autobiographical sketch he talks about his background and the serendipitous nature of discovery that does not always follow straightforward and obvious paths.

An Address delivered on the occasion of his Jubilee, 1891

In the course of the past year, and most recently on the occasion of the celebration of my seventieth birthday, and the subsequent festivities, I have been overloaded with honours, with marks of respect and of goodwill in a way which could never have been expected. My own sovereign, his Majesty the German Emperor, has raised me to the highest rank in the Civil Service; the Kings of Sweden and of Italy, my former sovereign, the Grand Duke of Baden, and the President of the French Republic, have conferred Grand Crosses on me; many academies, not only of science, but also of the fine arts, faculties, and learned societies spread over the whole world, from Tomsk to Melbourne, have sent me diplomas, and richly illuminated addresses, expressing in elevated language their recognition of my scientific endeavours, and their thanks for those endeavours, in terms which I cannot read without a feeling of shame. My native town, Potsdam, has conferred its freedom on

* Reprinted from Hermann von Helmoltz, "Autobiographical Sketch" In: *Popular Lectures on Scientific Subjects Second Series* (London: Longmans, Green, and Co., 1908) 266-291 Translated by E. Atkinson, PhD.; public domain

me. To all this must be added countless individuals, scientific and personal friends, pupils, and others personally unknown to me, who have sent their congratulations in telegrams and in letters.

But this is not all. You desire to make my name the banner, as it were, of a magnificent institution which, founded by lovers of science of all nations, is to encourage and promote scientific inquiry in all countries. Science and art are, indeed, at the present time the only remaining bond of peace between civilised nations. Their ever-increasing development is a common aim of all; is effected by the common work of all, and for the common good of all. A great and a sacred work! The founders even wish to devote their gift to the promotion of those branches of science which all my life I have pursued, and thus bring me, with my shortcomings, before future generations almost as an exemplar of scientific investigation. This is the proudest honour which you could confer upon me, in so much as you thereby show that I possess your unqualified favourable opinion. But it would border on presumption were I to accept it without a quiet expectation on my part that the judges of future centuries will not be influenced by considerations of personal favour.

My personal appearance even, you have had represented in marble by a master of the first rank, so that I shall appear to the present and to future generations in a more ideal form; and another master of the etching needle has ensured that faithful portraits of me shall be distributed among my contemporaries.

I cannot fail to remember that all you have done is an expression of the sincerest and warmest goodwill on your part, and that I am most deeply indebted to you for it.

I must, however, be excused if the first effect of these abundant honours is rather surprising and confusing to me than intelligible. My own consciousness does not justify me in putting a measure of the value of what I have tried to do, which would leave such a balance in my favour as you have drawn. I know how simply everything I have done has been brought about; how scientific methods worked out by my predecessors have naturally led to certain results, and how frequently a fortunate circumstance or a lucky accident has helped me. But the chief difference is this—that which I have seen slowly growing from small beginnings through months and years of toilsome and tentative work, all that suddenly starts before you like Pallas fully equipped from the head of Jupiter. A feeling of surprise has entered into your estimate, but not into mine. At times, and perhaps even frequently, my own estimate may possibly have been unduly lowered by the fatigue of the work, and by vexation about all kinds of futile steps which I had taken. My colleagues, as well as the public at large, estimate a scientific or artistic work according to the utility, the instruction, or the pleasure which it has afforded. An author is usually disposed to base his estimate on the labour it has cost him, and it is but seldom that both kinds of judgment agree. It can, on the other hand, be seen from incidental expressions of some of the most celebrated men, especially of artists, that they lay but small weight on productions which seem to us inimitable, compared with others which have been difficult, and yet which appear to readers and observers as much less successful. I need only mention Goethe, who once stated to Eckermann that he did not estimate his poetical works so highly as what he had done in the theory of colours.

The same may have happened to me, though in a more modest degree, if I may accept your assurances and those of the authors of the addresses which have reached me. Permit me, therefore, to give you a short account of the manner in which I have been led to the special direction of my work.

In my first seven years I was a delicate boy, for long confined to my room, and often even to bed; but, nevertheless I had a strong inclination towards occupation and mental activity. My parents busied themselves a good deal with me; picture books and games, especially with wooden blocks, filled up the rest of the time. Reading came pretty early, which, of course, greatly increased the range of my occupations. But a defect of my mental organisation showed itself almost as early, in that I had a bad memory for disconnected things. The first indication of this I consider to be the difficulty I had in distinguishing between left and right; afterwards, when at school I began with languages, I had greater difficulties than others in learning words, irregular grammatical forms, and peculiar terms of expression. History as then taught to us I could scarcely master. To learn prose by heart was martyrdom. This defect has, of course, only increased, and is a vexation of my mature age.

But when I possessed small mnemotechnical methods, or merely such as are afforded by the metre and rhyme of poetry, learning by heart, and the retention of what I had learnt, went on better. I easily remembered poems by great authors, but by no means so easily the somewhat artificial verses of authors of the second rank. I think that is probably due to the natural flow of thought in good poems, and I am inclined to think that in this connection is to be found an essential basis of aesthetic beauty. In the higher classes of the Gymnasium I could repeat some books of the Odyssey, a considerable number of the odes of Horace, and large stores of German poetry. In other directions I was just in the position of our older ancestors, who were not able to write, and hence expressed their laws and their history in verse, so as to learn them by heart.

That which a man does easily he usually does willingly; hence I was first of all a great admirer and lover of poetry. This inclination was encouraged by my father, who, while he had a strict sense of duty, was also of an enthusiastic disposition, impassioned for poetry, and particularly for the classic period of German Literature. He taught German in the upper classes of the Gymnasium, and read Homer with us. Under his guidance we did, alternately, themes in German prose and metrical exercises—poems as we called them. But even if most of us remained indifferent poets, we learned better in this way, than in any other I know of, how to express what we had to say in the most varied manner.

But the most perfect mnemotechnical help is a knowledge of the laws of phenomena. This I first got to know in geometry. From the time of my childish playing with wooden blocks, the relations of special proportions to each other were well known to me from actual perception. What sort of figures were produced when bodies of regular shape were laid against each other I knew well without much consideration. When I began the scientific study of geometry, all the facts which I had to learn were perfectly well known and familiar to me, much to the astonishment of my teachers. So far as I recollect, that came out incidentally in the elementary school attached to the Potsdam Training College, which

I attended up to my eighth year. Strict scientific methods, on the contrary, were new to me, and with their help I saw the difficulties disappear which had hindered me in other regions.

One thing was wanting in geometry; it dealt exclusively with abstract forms of space, and I delighted in complete reality. As I became bigger and stronger I went about with my father and my schoolfellows a great deal in the neighbourhood of my native town, Potsdam, and I acquired a great love of Nature. This is perhaps the reason why the first fragments of physics which I learned in the Gymnasium engrossed me much more closely than purely geometrical and algebraical studies. Here there was a copious and multifarious region, with the mighty fulness of Nature, to be brought under the dominion of a mentally apprehended law. And, in fact, that which first fascinated me was the intellectual mastery over Nature, which at first confronts us as so unfamiliar, by the logical force of law. But this, of course, soon led to the recognition that knowledge of natural processes was the magical key which places ascendency over Nature in the hands of its possessor. In this order of ideas I felt myself at home. I plunged then with great zeal and pleasure into the study of all the books on physics I found in my father's library. They were very old-fashioned; phlogiston still held sway, and galvanism had not grown beyond the voltaic pile. A young friend and myself tried, with our small means, all sorts of experiments about which we had read. The action of acids on our mothers' stores of linen we investigated thoroughly; we had otherwise but little success. Most successful was, perhaps, the construction of optical instruments by means of spectacle glasses, which were to be had in Potsdam, and a small botanical lens belonging to my father. The limitation of our means had at that time the value that I was compelled always to vary in all possible ways my plans for experiments, until I got them in a form in which I could carry them out. I must confess that many a time when the class was reading Cicero or Virgil, both of which I found very tedious, I was calculating under the desk the path of rays in a telescope, and I discovered, even at that time, some optical theorems, not ordinarily met with in text-books, but which I afterwards found useful in the construction of the ophthalmoscope.

Thus it happened that I entered upon that special line of study to which I have subsequently adhered, and which, in the conditions I have mentioned, grew into an absorbing impulse, amounting even to a passion. This impulse to dominate the actual world by acquiring an understanding of it, or what, I think, is only another expression for the same thing, to discover the causal connection of phenomena, has guided me through my whole life, and the strength of this impulse is possibly the reason why I found no satisfaction in apparent solutions of problems so long as I felt there were still obscure points in them.

And now I was to go to the university. Physics was at that time looked upon as an art by which a living could not be made. My parents were compelled to be very economical, and my father explained to me that he knew of no other way of helping me to the study of Physics, than by taking up the study of medicine into the bargain. I was by no means averse from the study of living Nature, and assented to this without much difficulty. Moreover, the only influential person in our family had been a medical man, the late Surgeon-General Mursinna; and this relationship was a recommendation in my favour among other applicants for admission to our Army Medical School, the Friedrich Wilhelms Institut, which very materially helped the poorer students in passing through their medical course.

In this study I came at once under the influence of a profound teacher—Johannes Müller; he who at the same time introduced E. Du Bois Reymond, E. Brücke, C. Ludwig, and Virchow to the study of anatomy and physiology. As respects the critical questions about the nature of life, Müller still struggled between the older—essentially the metaphysical—view and the naturalistic one, which was then being developed; but the conviction that nothing could replace the knowledge of facts forced itself upon him with increasing certainty, and it may be that his influence over his students was the greater because he still so struggled.

Young people are ready at once to attack the deepest problems, and thus I attacked the perplexing question of the nature of the vital force. Most physiologists had at that time adopted G.E. Stahl's way out of the difficulty, that while it is the physical and chemical forces of the organs and substances of the living body which act on it, there is an indwelling vital soul or vital force which could bind and loose the activity of these forces; that after death the free action of these forces produces decomposition, while during life their action is continually being controlled by the soul of life. I had a misgiving that there was something against nature in this explanation; but it took me a good deal of trouble to state my misgiving in the form of a definite question. I found ultimately, in the latter years of my career as a student, that Stahl's theory ascribed to every living body the nature a *perpetuum mobile* (perpetual motion). I was tolerably well acquainted with the controversies on this latter subject. In my school days I had heard it discussed by my father and our mathematical teachers, and while still a pupil of the Friedrich Wilhelms Institut I had helped in the library, and in my spare moments had looked through the works of Daniell, Bernouilli, D'Alembert, and other mathematicians of the last century. I thus came upon the question, 'What relations must exist between the various kinds of natural forces for a perpetual motion to be possible?' and the further one, 'Do those relations actually exist?' In my essay, 'On the Conservation of Force,' my aim was merely to give a critical investigation and arrangement of the facts for the benefit of physiologists.

I should have been quite prepared if the experts had ultimately said, 'We know all that. What is this young doctor thinking about, in considering himself called upon to explain it all to us so fully?' But, to my astonishment, the physical authorities with whom I came in contact took up the matter quite differently. They were inclined to deny the correctness of the law, and in the eager contest in which they were engaged against Hegel's Natural Philosophy were disposed to declare my essay to be a fantastical speculation. Jacobi, the mathematician, who recognised the connection of my line of thought with that of the mathematicians of the last century, was the only one who took an interest in my attempt, and protected me from being misconceived. On the other hand, I met with enthusiastic applause and practical help from my younger friends, and especially from E. Du Bois Reymond. These, then, soon brought over to my side the members of the recently formed Physical Society of Berlin. About Joule's researches on the same subject I knew at that time but little, and nothing at all of those of Robert Mayer.

Connected with this were a few smaller experimental researches on putrefaction and fermentation, in which I was able to furnish a proof, in opposition to Liebig's contention, that both were by no means purely chemical decompositions, spontaneously occurring, or brought about by the aid of the atmospheric oxygen; that alcoholic fermentation more especially was bound up with the presence of yeast spores which are only formed by reproduction. There was, further, my work on metabolism in muscular action, which afterwards was connected with that on the development of heat in muscular action; these being processes which were to be expected from the law of the conservation of force.

These researches were sufficient to direct upon me the attention of Johannes Müller as well as of the Prussian Ministry of Instruction, and to lead to my being called to Berlin as Brücke's successor, and immediately thereupon to the University of Königsberg. The Army medical authorities, with thankworthy liberality, very readily agreed to relieve me from the obligation to further military service, and thus made it possible for me to take up a scientific position.

In Königsberg I had to lecture on general pathology and physiology. A university professor undergoes a very valuable training in being compelled to lecture every year, on the whole range of his science, in such a manner that he convinces and satisfies the intelligent among his hearers—the leading men of the next generation. This necessity yielded me, first of all, two valuable results.

For in preparing my course of lectures, I hit directly on the possibility of the ophthalmoscope, and then on the plan of measuring the rate of propagation of excitation in the nerves.

The ophthalmoscope is, perhaps, the most popular of my scientific performances, but I have already related to the oculists how luck really played a comparatively more important part than my own merit. I had to explain to my hearers Brücke's theory of ocular illumination. In this, Brücke was actually within a hair's breadth of the invention of the ophthalmoscope. He had merely neglected to put the question, 'To what optical image do the rays belong, which come from the illuminated eye?' For the purpose he then had in view it was not necessary to propound this question. If he had put it, he was quite the man to answer it as quickly as I could, and the plan of the ophthalmoscope would have been given. I turned the problem about in various ways, to see how I could best explain it to my hearers, and I thereby hit upon the question I have mentioned. I knew well, from my medical studies, the difficulties which oculists had about the conditions then comprised under the name of Amaurosis, and I at once set about constructing the instrument by means of spectacle glasses and the glass used for microscope purposes. The instrument was at first difficult to use, and without an assured theoretical conviction that it must work, I might, perhaps, not have persevered. But in about a week I had the great joy of being the first who saw clearly before him a living human retina.

The construction of the ophthalmoscope had a very decisive influence on my position in the eyes of the world. From this time forward I met with the most willing recognition and readiness to meet my wishes on the part of the authorities and of my colleagues, so that for the future I was able to pursue far more freely the secret impulses of my desire for knowledge. I must, however, say that I ascribed my success in great measure to the circumstance that, possessing some geometrical capacity, and equipped with a knowledge of physics, I had, by good fortune, been thrown among medical men, where I found in physiology a virgin soil of great fertility; while, on the other hand, I was led by the consideration of the vital processes to questions and points of view which are usually foreign to pure mathematicians and physicists. Up to that time I had only been able to compare my mathematical abilities with those of my fellow-pupils and of my medical colleagues; that I was for the most part superior to them in this respect did not, perhaps, say very much. Moreover, mathematics was always regarded in the school as a branch of secondary rank. In Latin composition, on the contrary, which then decided the palm of victory, more than half my fellow-pupils were ahead of me.

In my own consciousness, my researches were simple logical applications of the experimental and mathematical methods developed in science, which by flight modifications could be easily adapted to the particular object in view. My colleagues and friends, who, like myself, had devoted themselves to the physical aspect of physiology, furnished results no less surprising.

But in the course of time matters could not remain in that stage. Problems which might be solved by known methods I had gradually to hand over-to the pupils in my laboratory, and for my own part turn to more difficult researches, where success was uncertain, where general methods left the investigator in the lurch, or where the method itself had to be worked out.

In those regions also which come nearer the boundaries of our knowledge I have succeeded in many things experimental and mechanical—I do not know if I may add philosophical. In respect of the former, like any one who has attacked many experimental problems, I had become a person of experience, who was acquainted with many plans and devices, and I had changed my youthful habit of considering things geometrically into a kind of mechanical mode of view. I felt, intuitively as it were, how strains and stresses were distributed in any mechanical arrangement, a faculty also met with in experienced mechanicians and machine constructors. But I had the advantage over them of being able to make complicated and specially important relations perspicuous, by means of theoretical analysis.

I have also been in a position to solve several mathematical physical problems, and some, indeed, on which the great mathematicians, since the time of Euler, had in vain occupied themselves; for example, questions as to vortex motion and the discontinuity of motion in liquids, the question as to the motion of sound at the open ends of organ pipes, &c. &c. But the pride which I might have felt about the final result in these cases was considerably lowered by my consciousness that I had only succeeded in solving such problems after many devious ways, by the gradually increasing generalisation of favourable examples, and by a series of fortunate guesses. I had to compare myself with an Alpine climber, who, not knowing the way, ascends slowly and with toil, and is often compelled to retrace his steps because his progress is stopped; sometimes by reasoning, and sometimes by accident, he hits upon traces of a fresh path, which again leads him a little further; and finally, when he has reached the goal, be finds to his annoyance a royal road on which he might have ridden up if he had been clever enough to find the right starting-point at the outset. In my memoirs I have, of course, not given the reader an account of my wanderings, but I have described the beaten path on which he can now reach the summit without trouble.

There are many people of narrow views, who greatly admire themselves, if once in a way, they have had a happy idea, or believe they have had one. An investigator, or an artist, who is continually having a great number of happy ideas, is undoubtedly a privileged being, and is recognised as a benefactor of humanity.

But who can count or measure such mental flashes? Who can follow the hidden tracts by which conceptions are connected?

That which man had never known,
Or had not thought out,
Through the labyrinth of mind
Wanders in the night.

I must say that those regions, in which we have not to rely on lucky accidents and ideas, have always been most agreeable to me, as fields of work.

But, as I have often been in the unpleasant position of having to wait for lucky ideas, I have had some experience as to when and where they came to me, which will perhaps be useful to others. They often steal into the line of thought without their importance being at first understood; then afterwards some accidental circumstance shows how and under what conditions they have originated; they are present, otherwise, without our knowing whence they came. In other cases they occur suddenly, without exertion, like an inspiration. As far as my experience goes, they never came at the desk or to a tired brain. I have always so turned my problem about in all directions that I could see in my mind its turns and complications, and run through them freely without writing them down. But to reach that stage was not usually possible without long preliminary work. Then, after the fatigue from this had passed away, an hour of perfect bodily repose and quiet comfort was necessary before the good ideas came. They often came actually in the morning on waking, as expressed in Goethe's words which I have quoted, and as Gauss also has remarked[13]. But, as I have stated in Heidelberg, they were usually apt to come when comfortably ascending woody hills in sunny weather. The smallest quantity of alcoholic drink seemed to frighten them away.

Such moments of fruitful thought were indeed very delightful, but not so the reverse, when the redeeming ideas did not come. For weeks or months I was gnawing at such a question until in my mind I was

Like to a beast upon a barren heath
Dragged in a circle by an evil spirit,
While all around are pleasant pastures green.

And, lastly, it was often a sharp attack of headache which released me from this strain, and set me free for other interests.

I have entered upon still another region to which I was led by investigation on perception and observation of the senses, namely, the theory of cognition. Just as a physicist has to examine the telescope and galvanometer with which he is working; has to get a clear conception of what he can attain with them, and how they may deceive him; so, too, it seemed to me necessary to investigate likewise the capabilities of our power of thought. Here, also, we were concerned only with a series of questions of fact about which definite answers could and must be given. We have distinct impressions of the senses, in consequence of which we know how to act. The success of the action usually agrees with that which was to have been anticipated, but sometimes also not, in what are called subjective impressions. These are all objective facts, the laws regulating which it will be possible to find. My principal result was that the impressions of the senses are only signs for the constitution of the external world, the interpretation of which must be learned by experience. The interest for questions of the theory of cognition, had been implanted in me in my youth, when I had often heard my father, who had retained a strong impression from Fichter's idealism, dispute with his colleagues who believed in Kant or He-

---

[13] Gauss, *Werke,* vol. v. p. 609."The law of induction discovered Jan. 23, 1835, at 7 A.M., before rising."

gel. Hitherto I have had but little reason to be proud about those investigations. For each one in my favour, I have had about ten opponents; and I have in particular aroused all the metaphysicians, even the materialistic one, and all people of hidden metaphysical tendencies. But the addresses of the last few days have revealed a host of friends whom as yet I did not know; so that in this respect also I am indebted to this festivity for pleasure and for fresh hope. Philosophy, it is true, has been for nearly three thousand years the battle-ground for the most violent differences of opinion, and it is not to be expected that these can be settled in the course of a single life.

I have wished to explain to you how the history of my scientific endeavours and successes, so far as they go, appears when looked at from my own point of view, and you will perhaps understand that I am surprised at the universal profusion of praise which you have poured out upon me. My successes have had primarily this value for my own estimate of myself, that they furnished a standard of what I might further attempt; but they have not, I hope, led me to self-admiration. I have often enough seen how injurious an exaggerated sense of self-importance may be for a scholar, and hence I have always taken great care not to fall a prey to this enemy. I well knew that a rigid self-criticism of my own work and my own capabilities was the protection and palladium against this fate. But it is only needful to keep the eyes open for what others can do, and what one cannot do oneself, to find there is no great danger; and, as regards my own work, I do not think I have ever corrected the last proof of a memoir without finding in the course of twenty-four hours a few points which I could have done better or more carefully.

As regards the thanks which you consider you owe me, I should be unjust if I said that the good of humanity appeared to me, from the outset, as the conscious object of my labours. It was, in fact, the special form of my desire for knowledge which impelled me and determined me, to employ in scientific research all the time which was not required by my official duties and by the care for my family. These two restrictions did not, indeed, require any essential deviation from the aims I was striving for. My office required me to make myself capable of delivering lectures in the

University; my family, that I should establish and maintain my reputation as an investigator. The State, which provided my maintenance, scientific appliances, and a great share of my free time, had, in my opinion, acquired thereby the right that I should communicate faithfully and completely to my fellow-citizens, and in a suitable form, that which I had discovered by its help.

The writing out of scientific investigations is usually a trouble-some affair; at any rate it has been so to me. Many parts of my memoirs I have rewritten five or six times, and have changed the order about until I was fairly satisfied. But the author has a great advantage in such a careful wording of his work. It compels him to make the severest criticism of each sentence and each conclu-sion, more thoroughly even than the lectures at the University which I have mentioned. I have never considered an investigation finished until it was formulated in writing, completely and without any logical deficiencies.

Those among my friends who were most conversant with the matter represented to my mind, my conscience as it were. I asked myself whether they would approve of it. They hovered before me as the embodiment of the scientific spirit of an ideal humanity, and furnished me with a standard.

In the first half of my life, when I had still to work for my ex-ternal position, I will not say that, along with a desire for knowledge and a feeling of duty as servant of the State, higher eth-ical motives were not also at work; it was, however, in any case dif-ficult to be certain of the reality of their existence so long as selfish motives were still existent. This is, perhaps, the case with all inves-tigators. But afterwards, when an assured position has been at-tained, when those who have no inner impulse towards science may quite cease their labours, a higher conception of their relation to humanity does influence those who continue to work. They gradually learn from their own experience how the thoughts which they have uttered, whether through literature or through oral in-struction, continue to act on their fellow-men, and possess, as it were, an independent life; how these thoughts, further worried out by their pupils, acquire a deeper significance and a more definite form, and, reacting on their originators, furnish them with fresh instruction. The ideas of an individual, which he himself has conceived, are of course more closely connected with his mental field of view than extraneous ones, and he feels more encouragement and satisfaction when he sees the latter more abundantly developed than the former. A kind of parental affection for such a mental child ultimately springs up, which leads him to care and to struggle for the furtherance of his mental offspring as he does for his real children.

But, at the same time, the whole intellectual world of civilised humanity presents itself to him as a continuous and spontaneously developing whole, the duration of which seems infinite as compared with that of a single individual. With his small contributions to the building up of science, he sees that he is in the service of something everlastingly sacred, with which he is connected by close bands of affection. His work thereby appears to him more sanctified. Anyone can, perhaps, apprehend this theoretically, but actual personal experience is doubtless necessary to develop this idea into a strong feeling.

The world, which is not apt to believe in ideal motives, calls this feeling love of fame. But there is a decisive criterion by which both kinds of sentiment can be discriminated. Ask the question if it is the same thing to you whether the results of investigation which you have obtained are recognised as belonging to you or not when there are no considerations of external advantage bound up with the answer to this question. The reply to it is easiest in the case of chiefs of laboratories. The teacher must usually furnish the fundamental idea of the research as well as a number of proposals for overcoming experimental difficulties, in which more or less ingenuity comes into play. All this passes as the work of the student, and ultimately appears in his name when the research is finished. Who can afterwards decide what one or the other has done? And how many teachers are there not who in this respect are devoid of any jealousy?

Thus, gentlemen, I have been in the happy position that, in freely following my own inclination, I have been led to researches for which you praise me, as having been useful and instructive. I am extremely fortunate that I am praised and honoured by my contemporaries, in so high a degree, for a course of work which is to me the most interesting I could pursue. But my contemporaries have afforded me great and essential help. Apart from the care for my own existence and that of my family, of which they have relieved me, and apart from the external means with which they have provided me, I have found in them a standard of the intellectual capacity of man; and by their sympathy for my work they have evoked in me a vivid conception of the universal mental life of humanity which has enabled me to see the value of my own researches in a higher light. In these circumstances, I can only regard as a free gift the thanks which you desire to accord to me, given unconditionally and without counting on any return.

21

On the Process of Becoming a Great Scientist Morgan C. Giddings

Morgan C. Giddings is a professor of biochemistry at Boise State University focusing on computational biology. Her research focus has been both in genomics and proteomics (understanding protein interactions and diversity). In this article she gives some tips about becoming a great scientist.

In the Editorial "Ten Simple Rules for Doing Your Best Research, According to Hamming" [1], Erren and colleagues discussed ten ideas originally presented by Hamming for how to do great science. I am grateful that the authors started this discussion. Scientific careers are very challenging, and there is a lack of training in many graduate programs to provide this kind of career meta-advice. Such discussions are a good starting point, and young scientists should take them seriously.

In the vein of promoting further debate and discussion, I provide here a different and perhaps deeper look at what makes a successful scientist. While I can't claim to have the reputation of Hamming, I grew up in a family of well-known scientists, and have had plenty of chances to observe the trajectories of scientific careers over my lifetime. Based on that experience, I propose the following as a somewhat distinct set of guidelines for doing the best research:

* Reprinted from Morgan C. Giddings,"On the Process of Becoming a Great Scientist", *PLoS Computational Biology* Vol. 4 No. 2: e33 (2008); Creative Commons Attribution License

1. Don't worry about age, worry about being exposed to new ideas.

While it appears that age plays a role in scientific creativity, it has not been well examined whether that role is biologically causative. There are many *social changes* that usually occur as anyone ages, which may play a greater role than biology does in the age-related creativity decline. Older scientists usually become boxed into their fields of expertise, and come to be seen as "experts." As such, they are less likely to have their ideas directly challenged by others, and less likely to be exposed to radically new ideas or different fields. I have seen many anecdotal references to Einstein's creative powers reducing as he aged, as his best work was done in his 20s. But this ignores a major factor: during his creative years, he was a patent clerk who was seen as a "nobody," whereas in his later years he was an eminent professor. Being a nobody has certain creative advantages—for one, there is not much to lose by promoting radical new ideas, because one has no reputation or established career at stake. Also, one is not expected to follow the "party line," regardless of the latest scientific fashion that happens to be in vogue.

Promoting new ideas can often be a minefield for one's career, since there is usually a long period of violent resistance to new ideas. Barry Marshall had to drink a culture of *H. pylori* to give himself an ulcer, in order to overcome resistance to the idea that this organism caused ulcers [2]. Now, more than 20 years later, he and co-discoverer Robin Warren have the Nobel Prize, and the role of *H. pylori* in ulcers is widely accepted.

In today's competitive grant world, this phenomenon is exacerbated. It is dangerous to one's funding to go against the trend, and if there is a lab to support and mouths to feed, the disincentives are great. This phenomenon stifles creativity, perhaps far more than biological age does.

If one is therefore concerned about retaining scientific creativity, perhaps the best solution is to force exposure to new ideas, concepts, and people. Hamming also discussed the importance of this kind of exposure by "keeping your door open." [3] I think that more than just keeping one's door open, a more direct way of doing this is to become involved in entirely new fields from time to time, which tends to promote creative thinking outside established dogma. So, don't worry about your age, worry about whether you are continuing to expose yourself to new and challenging ideas.

2. Tinker.

While it is not frequently acknowledged either in the popular press or in scientific literature, a significant fraction of scientific discovery is the result of serendipity (or to put it more bluntly, luck). From the discovery of penicillin by Fleming to the discovery of new ionization techniques such as MALDI that power modern mass-spectrometry based proteomic research, luck has frequently played a big role. Such discoveries are generally attributed to hard work and genius, rather than to luck. Doing so gives the "genius" too much credit and luck too little.

Often the big discoveries come from someone noticing an inconsistency or oddity in their surroundings or experiments, then doggedly working to figure out what is causing it. So perhaps being a great scientist is less about "genius" than it is about willingness to pursue the unusual at the expense of pursuing the usual. This comes back to the argument about age: often, once one has become entrenched in a paradigm, blindness to inconsistencies grows, and so it takes someone from outside of a field to point those out and pursue them.

This should be encouraging news for those of us who don't consider ourselves geniuses. The best way to promote scientific success may be to maximize exposure to chance occurrence and events—especially those that have more upside than downside potential. So, don't just ignore those little inconsistencies that arise in your work, give them some room for consideration. This is something anyone can do, though it takes time and courage (see point 3, below).

In addition, to be creative and remain open to fortuitous occurrences, the mind needs a rest from time to time. One can be buried in the lab 20 hours a day, and easily become lost in the self-created world where the little oddities begin to escape notice. Fleming discovered penicillin upon return from a long vacation, and his fresh mind may have contributed to the key observation he made on the effect of mold upon bacterial cultures. So it is critical to balance hard work with other activities, particularly those that provide exposure to new and different challenges: travel, sports, hobbies, family, or whatever.

3. Take risks.

Risk taking is where most of the big discoveries in science lie. Recall Dr. Marshall and *H. pylori*: he was willing to swallow a culture of the bacterium to prove his theory. And later, he shared the Nobel Prize for it. It may not be wise to go around drinking random bacterial cultures in the hopes of discovering something new. But it is important when something outside the current scientific fashion is discovered, to at least consider the risks and possible payoffs of pursuing it. Those who do pursue such ideas may find it hard to get funding for them. Others may say it is a bad idea. People may reject papers, expressing vehement opposition to a new idea. For really groundbreaking ideas, there may even be hecklers at talks! But, as Hamming pointed out in his lecture: "The great scientists, when an opportunity opens up, get after it and they pursue it." [3]

Pursuing new lines of inquiry can be very discouraging at times, but it is all part of the process any new idea goes through to transform from fringe to mainstream. I recall one major experience I had with this. Around 1996, I came up with an idea for doing DNA sequencing reactions in a test tube in a way that is very much like pyrosequencing today. After presenting it to a mentor and having it shot down, I gave up on it and went back to my "safe" work. While that was not a great time to pursue a new line of work outside my graduate studies, perhaps I should not have given up so quickly, considering the importance of pyrosequencers now.

Risk taking may be a particular challenge for female scientists. It seems that cultural norms discourage risk taking in young girls more so than in boys, and this can carry forward through to adulthood and into scientific careers. The top female scientists I know of take risks in their work, but they seem to be a minority. So it seems especially important for mentors of female students, postdocs, and young faculty, to provide encouragement in this regard. This same issue may apply to other minorities in science as well.

4. Enjoy your work!.

It is quite easy in today's science to get caught up in the "external rewards" game, meaning: seeking praise, high profile publications, and honors or awards. But these are transient and illusory rewards. The prestigious prizes and high profile publications are often a lottery—in addition to some of the factors above, there is a lot of luck involved in who happens upon the "really big" discoveries. One may or may not get lucky, and may or may not get recognition for that. Sometimes recognition only comes after the prime of one's career—John Fenn received the Nobel Prize at 85 years old. That's a long time to wait for reward if you're just doing science for the sake of such rewards (I doubt that was Fenn's motivation for discovering electrospray ionization).

A different and much more gratifying way to pursue a career is to simply enjoy the work! Do science for the sake of doing it. This is as likely as anything to lead to big discoveries and fame. But even if those things don't happen, you are enjoying yourself, and life is too short not to do so.

5. Learn to say "No!"

Over the span of a career, one gets asked to do many non-science activities: serving on committees, grant reviews, paper reviews, and so on. While it is important to contribute effort to these things to keep the system functioning, it is necessary to set a limit, so that they don't take over the fun of doing science itself. The system will not collapse just because one says "no" from time to time in order to preserve time to do science. Learning to say "no" is particularly important for young faculty, who find themselves barraged with such requests, and who can easily get sucked into full-time committee duties. It is wise to step back frequently and ask, "overall, is this work I am doing fun?" If the answer is no, perhaps it is time to revisit points 1 and 4 above, and consider diving into a new area.

6. Learn to enjoy the process of writing and presenting.

Note the distinction in this guideline from: "learn to write and present well." Many students I encounter dislike writing more than anything else they do. As a result, when it comes time to write a paper, it is a struggle from start to finish, both for them and for those working with them. When one doesn't like doing something, procrastination is the most common response. Procrastination and good writing don't mix. I say this even though I am someone who, as an undergraduate, would work all night on a term paper to turn it in at the last moment, and often receive an "A." But in the real world of scientific paper writing, that first draft just won't cut it. It usually takes three or more significant rewritings and lots of input from others to get it right. Combine that with procrastination and it's a recipe for not getting a good paper out in a timely fashion, or perhaps not at all.

So the key is to figure out how to enjoy the writing process, thereby encouraging oneself to avoid procrastination. There is no one formula that works for everyone—some people need utter peace and quiet for their writing. Others prefer writing at a coffee shop, or to have music playing. The thing is to figure out what works, and to stick with it, training oneself to have positive mental associations with writing.

Robert Boice, in his book *Advice for New Faculty Members*, suggests the key is to do a little bit of writing every day [4]. The goal is simply get the ideas on the page, without worrying about their form at the beginning. By doing this a little bit every day—perhaps only 30–60 minutes—it is amazing how quickly and enjoyably a big writing project can take shape through a process of gradual evolution.

This often takes significant retraining, however. Many of us begin with the notion that writing should come in sudden bursts of dramatic creation. This message is conveyed frequently in movies that portray an author writing a novel in a sudden last minute rush, and it is reinforced in high school and college by many of us learning to get away with writing papers at the last minute (and still doing well). Reprogramming that unrealistic expectation out of one's head is therefore a key to learning to enjoy writing.

The same principle applies to giving a good presentation: enjoy its making and giving. Forget everything you ever learned about giving dry, stuffy presentations (i.e., all those things in the document *How to Make a Scientific Lecture Unbearable*) [5]. While it is critical to have good science in your talk, it is equally critical to bring that science to life for the audience. That is nigh impossible if you are scared to death of being in front of the audience, or if you are completely bored by your subject matter. If you are bored, the audience will surely be bored, and you might as well not have wasted their time—or your own.

The last thing a reader or talk attendee wants to see is a bunch of data just to prove that you did some work. It is much more interesting to tell a story. The story begins with why you started the work in the first place (the big reasons, not just "because my advisor told me to"), it usually has mystery and intrigue (e.g., dead ends, which are worth reporting only if they helped lead you to the final answer), and some kind of dramatic conclusion (which challenges the audience to think about things in a new way). This may seem like overstatement, but having sat through many extraordinarily dry, boring scientific talks (and having read many dry papers), I find that the ones that stand out are those that have such elements. If there is a lack of enthusiasm for the work you are doing, that may be a sign that it's the wrong work for you to be doing.

It can be a fun challenge to figure out who your audience is and what they will respond to. For example, when I was a postdoctoral researcher, I once gave a group meeting presentation accompanied by sound effects borrowed from Monty Python. We all had a good laugh, and I still managed to convey some science, too. But I would never do this at a scientific conference. Yet at a conference with a series of 15 minute talks, it is still possible to give a presentation that stands out—by enjoying its making and giving, and fine-tuning it for that audience. Elements such as presenting clear, understandable slides, and providing adequate introduction and background to the audience are very important. But it is most important to discuss subject matter that you have enthusiasm about.

Once one has learned to enjoy writing and presenting, it is very likely that writing well and presenting well will follow, since it is more difficult to do a truly poor job of something one enjoys doing.

7. See the big picture and keep it in mind.

Understanding and conveying the big picture for one's work is perhaps the greatest challenge facing young scientists. It is difficult to make the transition from a life of undergraduate classwork—where every step is prescribed by the instructor—to the pursuit of authentic research in graduate school, where there is no a simple formula to follow to pursue a successful line of research. At the start of a research career, the subject matter is often prescribed by one's advisor, and as a result, it is very common for students to simply rely on the advisor's word that it is important work to be doing, without really thinking about it, in keeping with the earlier mode of operation from undergraduate days. This lack of introspection regarding the "why" translates into many problems down the road, including: bad presentations (because no motivation for the work is given), bad manuscripts (because no motivation for the work is given), and, often, bad morale (because one comes to feel like a robot turning a crank).

From the start, it is critical to be very familiar with the Why. Why are you doing the work? Who will care about it, either now, or in the future? Is it likely to have any benefit? Note that the answers to these questions are often not easy. Many times discoveries are made long before they are ever put to practical use, and that use is often well outside the vision of the originator. So the key to this point is to think about the Why, even if there is no simple answer. Another way of stating this is that there should be some explicit and stated motivation for the work, even if it is just "intellectual curiosity." That kind of introspection will help with one's own motivation in doing the work, and just as importantly, this will translate into better presentations and papers (because of making it more fun, as discussed in point 6 above).

It would be gratifying to come up with guidelines 8–10 just for numerical conformity. However, lacking an additional and meaningful guideline to give, instead I would restate one in particular: guideline 4, enjoy your work. All of the great scientists I have encountered are those who really enjoy what they are doing.

The astute reader may notice that most of the above rules are about process, rather than end result. This is to counter a phenomenon endemic to our culture: results count, and so advice is usually tailored to how to get those results in the quickest and most obvious manner. However, by attempting to short-circuit the thinking about process, in order to achieve the quickest result, often the end result is not a better one, and more importantly, leads to little long-term gratification.

An example is the advice to "work hard." While one who works hard is usually more productive than one who doesn't, working too hard can be counterproductive. The rule could instead be stated "work hard enough," but then the question becomes: how much work is "hard enough?" That leads to a quagmire of endless debate about how much work might make one most productive (and even how said productivity is measured—is it citations, prestigious prizes, grant money, salary, or…?).

If one focuses instead on the processes involved in doing science, then the answers to such questions are much more obvious. Enough work is exactly the amount at which one can maintain enjoyment of the process of work, without burning out (which is not enjoyable) or becoming socially isolated (which is not enjoyable). If that amount of work is not enough to maintain a scientific career, then a different career may need to be considered, where such enjoyment can be found. Because, in the end, one may have many medals or honors bestowed, but those are transient scraps of paper or metal. True satisfaction with doing something worthwhile lasts for a lifetime.

Acknowledgments

The author would like to thank Mark Holmes, Adrienne Cox, and Jameson Miller for a critical reading of this article.

References

[1] Erren, T.C., Cullen, P., Erren, M., & Bourne, P.E. "Ten simple rules for doing your best research, according to Hamming." *PLoS Computational Biology,* Vol. 3, e213 (2007).

[2] Marshall, B. "Helicobacter connections." *Chem Med Chem* Vol. 1, 783 (2006).

[3] Hamming, R., "You and your research." In: Ed. Kaiser, J.F., Transcription of the Bell Communications Research Colloquium Seminar; 7 March 1986; Morristown, New Jersey, United States.

[4] Boice, R., *Advice for new faculty members.* (Needham Heights, MA: Allyn and Bacon, 2000).

[5] Kohn, A., "How to make a scientific lecture unbearable." *Annals of Improbable Research.* Available from http://www.improbable.com (2003).

22

The Beginner's Guide to Winning the Nobel Prize: A Life in Science Chapter Nine: How to Win a Nobel Prize

Peter Doherty

Peter Doherty is an Australian veterinary surgeon and one of the pioneers in research on the immune system and its recognition of antigens. He won the Nobel Prize in Physiology or Medicine in 1996 with Rolf Zinkernagel for their discoveries of how T-Cells use proteins from a structure called the Major Histocompatibility Complex (MHC) to recognize foreign antigens and provoke an immune response. In his autobiography, The Beginner's Guide to Winning the Nobel Prize, the final chapter is advice on how to have a successful career in research.

So you want to win a Nobel Prize: to become famous, powerful and maybe even very wealthy? If that's your ambition I can't help you. There is no instruction manual or course that can guide you to a Nobel Prize and, numerically speaking, most of us have more chance of winning an Olympic gold medal. There's also another difference: an Olympic medallist might go on to win a Nobel, but can you imagine Albert Einstein or Bertrand Russell competing in the decathlon? I was brutally reminded of this when I had to present a large cheque to Michael Chang for winning the St Jude Tennis Classic in Memphis. We were both winners in one sense or another but, though Michael might conceivably change his life at some stage to become a great scientist or writer, there is

* Reprinted with permission of the author from Peter Doherty, *The Beginner's Guide to Winning the Nobel Prize* (New York: Columbia University Press, 2006) 238-253.

no way that I could ever beat even an 85-year-old Chang or Sampras on the court.

Now that I've had your attention and you have read this far, I hope you will recognise something of what it takes to make an outstanding research scientist. It involves a personal recognition that humanity advances by insight, discovery and a capacity for serious effort and commitment. So, following what I've written here won't guarantee a trip to Stockholm or Oslo, but, with a little luck, it could lead to something worthwhile.

Try to solve major problems and make really big discoveries

The individual who is well educated, works enormously hard and has inherited extraordinary ability and intellectual capacity might just conceivably be able to identify a major problem at the Nobel level of achievement, then move ahead to solve it. From my experience such people are pretty rare, and may well be either alien life forms or the next stage in human evolution. Discovery is different. Nobody can decide to discover something, but there are ways of making a discovery more likely. Focus on generating new information and insights and look for unexpected outcomes and results. Accept nothing at face value and get in the habit of thinking unconventionally. Work hard, work smart and, with a bit of luck, serendipity will play its part.

Be realistic and play to your strengths

A trained veterinary surgeon like me knows, like all punters, that there are horses for courses. Everyone has to find out what sort of horse they are. Anyone with a brain that does best at ploughing long, straight furrows should give up on the idea of being an intellectual polo pony or steeplechaser. Perhaps a molecular biologist or organic chemist can also be a poet, but it's likely that most will do a lot better at one than the other. Science at its best is for people who love to ask questions and are delighted by discoveries that overturn established ideas and prejudices. If they have to choose between authority and evidence, basic scientists will always go with the evidence. Most scientists are notoriously contemptuous of authoritarian politics, for example. Any love affair between science and politics is always fraught with potential conflict, though the passion and betrayal that characterises tempestuous affairs often makes the best theatre, or press—as is usually the case.

Acquire the basic skills, and work with the right people

The elements that make an exceptional humanitarian or writer can be as varied as the individuals themselves—apart from the obvious ones, like a keen intellect and a serious sense of commitment. On the other hand, scientists have an absolute need for in-depth, specialist training at the university undergraduate level and beyond. Though it's not essential, it helps to be born into an intellectual and supportive family, grow up in the United States, Europe, Japan, Canada or Australia, attend an academic school and a great university, and train with a top person. Some aspirants try very hard to work with a Nobel laureate, as they have the right to nominate people for Nobel Prizes—and only those who are nominated are considered. Senior scientists like to think that they create enduring 'schools', so it can help to be part of such a lineage. None of these factors, however, will guarantee a Nobel. Sometimes the idiosyncratic outsider will rise to the top over those within the big tent of mutual reinforcement, where it can be too warm and too comfortable. Thank goodness for that. Otherwise science would be a stuffy and obsequious business.

Learn to write clearly and concisely

Many people who are very good at science are great doers, but uninspired writers. It isn't necessary to be a Shakespeare or a Michael Ondaatje, but anyone who wants to be recognised as a top scientist must be able to write clear, concise English. English is the language of science and many countries, among them Malaysia and Singapore, that are building their science profiles teach in English at school and university level. Science is about telling good, readable, memorable stories.

Work in an appropriate field

Nobel Prizes recognise some, but by no means every, aspect of what we may think of as the high culture of humanity. There are no Nobel Prizes for the visual arts, for music or for dance, so these might be fields you would want to avoid. The performing arts have been recognised only twice, to my knowledge, by the 1997 Literature award to the Italian playwright Dario Fo and the 1953 Literature Prize to the British leader Winston Churchill, who was, of course, a noted orator as well as a writer—although the public advocacy required of many Peace laureates as they seek to promote their particular interests might be considered partly under the head of 'performance.' For scientists, speaking about their work to both specialist and broader groups is an essential component of receiving credit and building a reputation. You and your work will remain anonymous if you just stay home. You will also need to exercise caution in the area of science you work in. While some areas of research are not specifically identified as targets for Nobel Prizes, they may slip under the wire in another category. Though pure mathematics may have been excluded, theory based in mathematics is clearly central to physics and economics, as exemplified in the Economics awards to John Nash in 1994 and James Mirlees in 1996. Geology, for instance, is not specifically identified, but it is possible that someone who trained primarily as a geologist might be honoured for contributions to physics or chemistry. There is no agriculture prize, but plant scientists who have been recognised include the wheat breeder Norman Borlaug referred to earlier, who won the Peace Prize in 1970 for the part he played in what has become known as the green revolution; the agricultural scientist and research administrator John Boyd Orr was awarded the Peace Prize in 1949 for establishing the FAO; and the plant geneticist Barbara McClintock the Prize for Medicine in 1983 for jumping genes in corn. Still, anyone who is set on the idea of a Nobel Prize would probably leave the plants to someone else. Otherwise, there is a world food prize.

Find and cultivate your true passion

Despite everything I've just said in the above, one of the best things that can happen in life is to discover a line of enquiry that really grabs your interest. Someone who has found a passion that doesn't fit the Nobel, or any other mould that conventionally leads to the 'glittering prizes', should forget the award and go for the satisfaction and the excitement of what they love doing, whether it's philosophy or building surf boards. That's where the prizes that really matter are to be found. According to the poet Ezra Pound, who would certainly have been ruled out of consideration for a Nobel Prize because of his fascination with Mussolini's fascism, 'What thou lovest well remains, the rest is dross'. Preoccupation with dross and irrelevance is a sure-fire way to avoid the Nobel Prize. Most of those who do win have not only achieved over a long period, but are likely to have given their full attention, energy and enthusiasm to what they do. On the other hand, a passion for working out mechanisms for usefully recycling the dross of packaging, junked cars and so forth that we produce in our daily lives could lead to one of the numerous environment prizes that can be identified by searching the web.

Focus and don't be a dilettante

Most scientists and economists are identified with a particular sub-field of investigation for much of their lives. Sometimes the best scientists will take on new challenges, but the majority—and particularly the prize winners—tend to remain within the same broad field, like cancer biology, neurobiology or immunology, where they are well known and regarded. Bright people who hop around from one topic to another often achieve very little. The exception may be the Nobel Peace Prizes. Concentration of effort may be required only in the relative short term for the humanitarian who wins a Peace Prize—the successful resolution of a major confrontation, for example, can depend on an individual's political power or stature as a negotiator. These credentials are likely to have been achieved in a completely different context, such as being US Secretary of State: Cordell Hull, Peace Prize, 1945; Henry Kissinger, Peace Prize, 1973. Other causes, like the elimination of landmines, gain traction only because someone like the 1997 Peace laureate, Jody Williams, is totally dedicated to finding a solution. The novelist or the poet may contribute in a variety of forms, though their style and approach may be consistent. Writing itself is the most intense of human activities. Again, it's all about intensity and commitment.

Be selective about where you work

As a scientist, your chances of achieving anything can be greatly diminished by working in an institution that is under-resourced financially, does not value creativity or demoralises even the bright people that it manages to recruit. The places that nurture winners don't all look the same, and can vary from small private institutions like CalTech, to massive Ivy League conglomerates like Harvard University, to state institutions like Southwestern Medical School in Dallas. Every one is different, so find an environment that suits your personality and work habits. Being in the regular company of colleagues who are stimulating to talk to and living in a culture that values creativity and insight contribute mightily to a satisfying life, even if the big prize doesn't come your way.

Value evidence and learn to see what's in front of your nose

In the end, science is about data and being able to 'read' the real meaning of what you find. Keep an open mind, be prepared to think laterally, and be instructed by nature and observation. Great literature and visionary science share a characteristic: the reader recognises that there is a sense of truth. The resolution of conflict or the navigation of an impasse that has been achieved by a Nobel Peace laureate also reflects a capacity to recognise, then act on the underlying reality. Such people bring to the task not only a substantial intellect, but also a clear perception of what can be achieved.

Think outside the box

Following the obvious path is not likely to lead to a novel question, interpretation or solution. If the way is both straight and narrow, the odds are that somebody else already will have gone down that road. The mind works in strange ways, and it can help to short-circuit, or bypass, normal thought processes. Edward de Bono formalised one such technique when he identified the 'lateral thinking' approach. When struggling with a scientific problem, it often helps to 'draw' the possibilities, either in your mind or on a piece of paper. Human beings think in both words and pictures. Illuminating ideas come at odd times, in the shower, for instance, or on the top of a mountain. Ilya Mechnikov, who won the Medicine Prize in 1908, famously discovered phagocytosis when, bored and at the beach, he poked small thorns into starfish larvae and watched the inflammatory cells congregate at the site of injury. Get rid of the clutter, and let the mind roam.

Physical activity, even if it's only walking, can also work to free the thought-processes. New ideas often seem to pop up when the mind is idling or half-concentrating on some more mechanistic activity, like gardening or building a bicycle shed. The 1993 Chemistry Prize winner, Kary Mullis, describes in his autobiography *Dancing Naked in the Mind Field* how the idea of the polymerase chain reaction came suddenly when he was tired and driving alone at night. This won't happen to you if your brain isn't grinding away at the problem in the background, which only happens if you are very attached to the particular question—obsessed even. Ask anyone who is married to a serious scientist, and you're likely to get into a discussion on the nature of obsession.

Talk about the problem

Don't be a lone wolf. Two heads are often better than one. The most obvious way to develop novel insights is to talk with others, particularly those who come at the issue from different backgrounds. Jim Watson's little book *The Double Helix* gives a strong sense of the intense interaction at Cambridge's Cavendish laboratory between the biologist (Watson) and the physicist Francis Crick (Medicine, 1962) as they tried to build their DNA model. The essential information that they had been using the wrong tautomeric form of the DNA bases came from Watson's chance conversation with a visiting American, the crystallographer Jerry Donohue. Without knowing this they were stuck. Though Rosalind Franklin took the key X-ray pictures that provided the solution for Watson and Crick, she and her King's College, London, colleague Maurice Wilkins failed to develop a rapport.

Rosalind remained very isolated and she did not solve the DNA problem.

There is a caveat. If you make a major discovery that could easily be repeated by others, it's best to keep quiet about it until the initial research report is either published or 'in press' in a top journal. Anyone who works in basic science at a high level is likely to have had the experience of the odd elliptical discussion with a colleague who is brimming with excitement about a new discovery but just can't afford to talk about it pre-publication, the I-could-tell-you-but-I-would-have-to-kill-you scenario. This is one time where you need to think like an MI5 operative. Australia is different now, but Rolf Zinkernagel and I benefited back in the 1970s from the isolation of this 'distant shore'. Talking openly in the local immunology discussion group, the weekly 'Bible class' run by the department head, Gordon Ada, certainly helped to clarify our thinking. That would be much riskier today: Australia is definitely in the scientific loop, and national and global communication means people can hear about a discovery almost the instant it is made. An inadvertent comment in someone's e-mail could provide the necessary clue for a competitor. Once our first papers (which are in the appendixes) were accepted for publication, however, we both did tours of the northern hemisphere and gave an enormous number of seminars to publicise our findings. I was exhausted by the time we met up for a final week at the 2nd International Congress of Immunology in Brighton, but Rolf was still jumping up and inserting the story of our discovery at every appropriate (and inappropriate) opportunity. Someone dubbed him 'HyperZink'. The sooner a major new scientific finding is out in the open the better, both for the discoverer(s) and for the field.

Tell the truth

Telling the truth about data is an absolute requirement in science. Apart from being the ultimate betrayal of scientific ethics, a lie can set everyone on the wrong course—including the perpetrator. The public revelation of such deception is likely to destroy a career. Those who withhold credit may get away with it for a time, but they run a risk if they want to be recognised ultimately as luminaries in their field.

Be generous and culturally aware

Freely acknowledging the achievements of others is a sure sign of someone who is confident of their own worth and integrity. Give credit where it is due, and acknowledge the work that came before yours and made your discovery possible. No senior scientist is ever hurt by giving precedence in authorship to junior colleagues who have done much of the hands-on work for a research paper.

Making virulent personal attacks on others in public, especially if the person concerned is young and inexperienced, can ultimately be counter-productive. Even if he or she is guilty of sloppy thinking or poorly presented data, it's easy enough to take someone aside later and talk through their conclusions. If they're obtuse and dogmatic, they will soon disappear from science anyway. Science is largely self-correcting, though this can take a while.

Be aware that some cultures are much more attuned than others to receiving (and returning) very direct and even harsh commentary. The type of intellectual sword play and blood-letting characteristic of an Oxbridge common room or high table can, for instance, seem both arrogant and vicious to those who live in the more polite, but concealed, world of US academia. Remember that, as George Bernard Shaw observed, "England and America are divided by a common language." Things may seem more familiar than they actually are. It's also important to know the subculture. Australian scientists are generally much more accustomed to tough criticism than the population at large, who may be inclined to punch you on the nose if you're too direct.

When it comes to awards, prizes, election to national academies and the like, it is probably important to have as few enemies as possible. Awards are neither gifts at the disposal of the gods on Mount Olympus nor the inevitable outcome of some automatic process of merit recognition. Selection committees consist of people who are broadly involved in the particular area of interest. At least some committee members will know a lot more about the candidate than the information that appears in the letters of recommendation and the published list of achievements. A strong, convincing and committed opponent can take down almost anyone.

Though it is likely that a truly spectacular discovery or body of achievement will ultimately be recognised, no matter what the personal characteristics of the recipient, there are situations where the decision could go either way. A record of vicious behaviour, or a suggestion that the individual concerned claims undue priority, can seriously damage a case. With the Nobel Prizes, for example, the decision can go to one field or another. There will always be alternative 'tickets'.

Be persistent and tenacious, but be prepared to fail

The old 'Protestant work ethic' is true for science: nothing worthwhile is likely to be easy. If at first you don't succeed, try, try, try again. Sayings like this pretty much describe the scientific life. My 8th-grade teacher, Miss Thompson, drummed into us: "Good, better best, may you never rest, till your good is better, and your better best." Maybe Miss T set some up for a life of deep neurosis, but anyone who wants to do experimental science has to be emotionally resilient. Murphy's Law—'Anything that can go wrong will go wrong'—certainly rules in experimental biology. The vital thing is to identify the problem, not apportion blame to one or other member of the research group (unless, of course, it is clearly justified). The Murphy experience, nevertheless, must be a hell of a lot worse for the 'big guys' with something like losing a Mars probe.

Every serious biomedical scientist will have had the experience of 'reverse alchemy', seeing what first looks like gold turn slowly into lead: an apparent breakthrough turns out to be a false trail that just can't be repeated in subsequent experiments. The process of moving systematically from high to low over an interval of weeks to months can be summarised by the Latin, *sic transit gloria*, thus passes glory. When this happens to you—and it invariably will—see it as a good time to break the cycle, stop the study and go to the pub or on vacation.

People who can't deal with failure, or can't acknowledge to themselves that they have been wrong, should probably avoid a life based in research. Nonetheless, it can be the case that those who are the most creative live on some sort of psychological edge. Such individuals have to develop strategies for dealing with the inevitable downs if they want to work in experimental biology. Sometimes they start well, but just can't continue. Most scientists who've been involved in leading a research effort over the longterm have had to deal with this type of tragedy.

Your time is precious

Winners and high-achievers will tell you that time is your greatest asset. It's accepted that novelists, painters and poets can be precious about protecting their creative time and space, but this territory isn't so clearly defined for those who do research for a living. Scientists work in large organisations and belong to global communities that organise meetings, and national and international societies. These things take up time! They are essential activities, and it is important that such roles should not be left only to those who are at the end of their careers. Even so, it is also necessary to be judicious and to set definite limits. 'Death by committees' is a particular trap for women scientists, who often have great negotiating skills but can be effectively drained by the demands of commitments, which may arise from the need for gender balance on this or that committee. That's fine for someone headed towards a career in academic administration, but such a commitment should be a conscious decision. When those committees come calling, just learn to say no.

Avoid prestigious administrative roles

Those bright people who accept a role as director, dean or president early in a career may well rule themselves out of the top league in the awards game. Leading a major institution is, of course, an energy-consuming activity. The same is true for running a major research program. My personal sense is that smart human beings commit themselves to what they like doing best. Some Nobel Prizes for experimental physics have gone to intellectually incisive, top administrators, but that isn't true for most areas of science. On the other hand, administrators earn the top dollars in the academic hierarchy, and those skills are increasingly in demand.

People who accept such posts sometimes win a Nobel Prize for work done earlier in their careers. Others who make the trip to Stockholm before they're too ancient often go on to be very effective university presidents or directors of prestigious research organisations. However, it's also the case that many who have the creativity and insight to succeed in research lack the types of skills and commitment that go into making an outstanding administrator. As Polonius in Hamlet would have it: "This above all, to thine own self be true."

Take care of yourself and live a long time

Given the nature of the Peace Prizes, a summons to Oslo is likely to come fairly soon after the achievement that is being recognised, but writers, scientists and economists may need to hang in there. It can take fifty years from the point of making a big discovery to the time that a Nobel committee comes to a decision that, at least from your point of view, is exactly the right one. Good habits start early: eat and drink moderately, take vacations, don't smoke or overuse recreational drugs (alcohol included), take regular exercise, avoid extreme sports, and seek professional help for suicidal thoughts. Scientists are a varied bunch and I know of very talented and effective individuals who have been taken out by each of the above factors. Given a way of life that is often more solitary, creative writers are likely to be even less armoured against such dangers.

Have fun, behave like a winner

See all the above.

On a more serious note, I want to emphasise that, while doing the type of work that leads to Nobel Prizes inevitably has its low points, on the whole it offers immense fulfilment. There can surely be no better feeling than the sense of having achieved an experimental result, or written a novel, poem or scientific paper that is personally satisfying, substantial and accessible, and is out there for the scrutiny of others. Nothing offers greater intellectual excitement than discovering something that no human being can ever have known before. Being able to live in a way that combines work with at least a measure of creativity is an immense privilege. My continued involvement in experimental science reflects this passion for discovery. Only a lunatic would expect to win a second Nobel Prize, so that certainly isn't the motive.

This is ultimately what science is about. Like most scientists I work in big institutions where, increasingly and inevitably, I don't know all the players, especially the younger ones, though they recognise me. If, for instance, I ride up alone in an elevator with some young postdoc whom I may never have seen before, I ask: "How is it going, and what are you doing?" The invariable experience is that they are delighted to be asked and are just bursting to summarise their particular science story. For those with the right mix of curiosity and commitment, the sense of probing a difficult question, of uncovering some basic—though maybe small—truth gives the greatest possible satisfaction. It isn't for everyone, but for those who get the message this is a good and honest way to live.

23

Excerpt about Productivity and Age

Robert Scott Root-Bernstein, Maurine Bernstein & Helen Garnier

Maurine Bernstein was a psychologist at UCLA who specialized in studying patterns of work among families and scientists. Helen Garnier is a statistical consultant for the Learning Research and Development Center, University of Pittsburgh, and the Institute of Cognitive Science, University of Colorado at Boulder. She is also a senior researcher on the Tying Words to Images of Science Teaching project, BSCS Center for Professional Development. Her research focus is education in reading, science, and mathematics. In this article they discuss how scientists maintain creative thinking as they age, even after previous big discoveries.

...In this study and in the historical examples cited, staying within one's area of expertise after having made a major contribution is *negatively correlated* with making major research contributions subsequently. No scientist who remained in his specialty after publishing two high-impact papers succeeded in publishing a third. Only those who changed fields contributed more high-impact papers.

Anecdotal evidence both from historical sources and from interviews with the scientists in the present study suggest why a diverse research program may be particularly fecund in the long-

* Reprinted with permission from Robert S. Root-Bernstein, Maurine Bernstein & Helen Gamier, "Identification of scientists making long - term, high-impact contributions, with notes on their methods of working," *Creativity Research Journal* Vol. 6 339-340 (1993).

term. For example, Nobel laureate Albert Szent-Györgyi wrote about the difficulty of remaining creative.

> *When I saw actomyosin for the first time [one of the discoveries for which he was awarded the Nobel Prize], I was convinced that in a fortnight I would understand muscle completely. Then I worked 20 years more without learning a thing…*[1]

Similarly, Hans Selye commented:

> *As the years went by, I managed to acquire every available facility that modern science can offer in the way of the most up-to-date techniques of histology, chemistry, and pharmacology. I have been given the means to construct one of the best equipped institutes of experimental medicine and surgery in the world and have acquired a staff of 53 trained assistants, technicians, and secretaries. Yet today as I look back upon those early observations in 1936, I am ashamed to say that, despite all this help, I have never again been able to add anything comparable in its significance to those first primitive experiments.* [2]

The problem, Szent-Györgyi said, is that

> *if one works for ten or twenty years on something, one needs a change of atmosphere. One gets stale; one doesn't see things* [1].

"Once a man has missed the solution to a problem when he passes it by," said Leo Szilard, "it is less likely he will find it next time." [1]

As the long-term, high-impact scientists in this study made clear in their interviews, lifelong producers of breakthroughs are not content simply to rework old fields or to refine prior insights or to become administrators of other people's research. They desire more and therefore change fields periodically. Several did this as an explicit aid to inventiveness. As one man (no. 18) said: "My advice to people whose research productivity is diminishing is to change fields." Several other long-term contributors (e.g., nos. 25 & 35) echoed his advice.

It was also noticed that those most likely to change fields are those who are constantly exploring other research problems even as they focus on one or two major ones. Two aspects of this phenomenon are significant. First, several studies have shown that scientists rarely (less than 10% of the time) can make significant headway on a problem by a direct, prolonged attack on it. Most report that they must abandon a problem before the solution occurs to them, or they find that the solution only arrives as a result of addressing another, related problem [3-5]. Thus, keeping several research problems going at once may benefit long-term, high-impact scientists by creating the best mental conditions for a high rate of insights [6]. When one project is going poorly, another may be going well, and in the meantime, the scientist may have an insight concerning the first. In some cases, the scientists (e.g., nos. 21 & 25) reported that they kept problems "simmering on a back burner" until adequate data, a new technique, or some insight finally made them accessible.

The other aspect of constantly exploring a range of research problems is more obvious. By trying many things, long-term, high impact scientists optimize the probability of finding new, significant, and important problems. Several of the scientists stated that they use strategies similar to that of Linus Pauling, who wrote that a scientist must "have lots of ideas and throw away the bad ones. And I think that this is part of it: that you aren't going to have good ideas unless you have lots of ideas and some sort of principle of selection." [7] The constant experimentation with new fields exhibited by the long-term, high-impact researchers would indicate the validity of Pauling's insight. Studies of both industrial chemists and university biomedical researchers confirm that the most effective and creative scientists are those who combined several specialties or technical functions as part of their normal work habits [6,8,9]. Nobel laureates Murray Gell-Mann, David Baltimore, Herbert Simon, and Vassily Leontief likewise agree that mastery of a wide range of tools and ideas from a diversity of weakly related disciplines combined with an understanding of emerging forms of mathematics are the keys to creating the sciences of the future [10].

Both the notion of exploring synergistic research areas and optimizing the probability of success by pursuing diverse interests are highly reminiscent of Gruber's [11-12] notion of *networks of enterprise*. Gruber defined such networks as consisting of a person's organization of purpose or definition of his or her working self; a structure that organizes what may appear to be a bewildering miscellany of activities; an organization of goals that provide different levels of risk and reward at different levels of aspiration to fit different changing moods and needs; and finally a sense of what makes a person's work individual and unique [11-14]. It is clear from the results of this study that the most successful scientists in the group developed networks of enterprise as complex and varied as those Gruber described for Charles Darwin and some of the other scientists he has studied.

Finally, consider the question of age. A wide variety of reports in the literature, including comments by the scientists, the opinions of many members of the scientific community, and a number of formal studies, suggest that scientific creativity declines irreversibly with age. Novices—those under the age of 35 or 40, make the majority of breakthroughs in science [15-21]. One reason often suggested for this phenomenon is that older scientists get saddled with increasing obligations—speaking engagements, administrative work, committees, fund raising, reviewing—that keep them from active research [22]. On this point, it is interesting to note that only two of the long-term, high-impact scientists (nos. 18 & 35) tried administrative work. One immediately abandoned it, stating that it was unfulfilling. The other remained as head of his department for 5 years despite his vehement protests. Eventually a medical leave forced the department to replace him and he went back to research for the rest of his career. In contrast, all of the short-term, high-impact scientists spent most of their later careers as administrators. Whether this change in career emphasis resulted from a lack of new ideas or contributed to it is not clear. One study, however, reported that industrial chemists producing many patents rarely manifested an interest in administrative work, whereas those producing no patents *always* manifested such an interest [23].

Another reason given for decline in creativity is the scientists inability to imagine making another discovery. As one of the short-term, high-impact scientists (no. 15) suggested,

> *suppose you do something great as a young man: So they're all going to say, "He ought to do another thing like this." He knows that in a lifetime, his chance of doing this again—the equivalent of this—is almost zero—so he's going to be a little depressed by this, isn't he?*

Setting impossible standards for oneself apparently interferes with ever trying anything again. Yet another reason sometimes given for decline in scientific creativity is that it is physiologically or mentally "inevitable." J.Z. Young, the famous neuroanatomist, wrote that:

> *There seems to be a limit beyond which new patterns and new connections are no longer easily formed. As we grow older the randomness of the brain becomes gradually used up. The brain ceases to be able to profit from experiment, it becomes set in patterns of laws. The well-established laws of a well-trained person may continue to be usefully applied to situations already experienced, though they fail to meet new ones. Here we see with startling clearness the basis of some of the most familiar features of human society: the adventure, subversiveness, inventiveness, and resource of the young; the informed and responsible wisdom of the old.* [24] p. 101

Although a distinction between intelligence and creativity may be necessary, it is important to note that the Miller Analogies Test scores of the scientists were constant (within experimental error) across the test period, and even into retirement. Thus, intelligence in the scientists did not seem to decline with age. Furthermore, the long-term, high-impact scientists demonstrated that scientific creativity need not decline. But note carefully the apparent reason: These men purposely placed themselves in the position of becoming novices again every 5 or 10 years. In effect they become mentally young by starting over again.

As several of them said in interviews, starting over again takes courage. It also apparently requires a different network of enterprise than the monolithic or monomaniacal style that characterizes one-time discoverers. Perhaps it is the courage to be ignorant again that fails most scientists as they grow older and not a matter of succumbing to physiological fatigue or set patterns of thought. Be that as it may be, it is evident that the *novice effect*, as it has been called [5, 25], worked for this select group of scientists. The novice effect may explain [18, 19] observation that three factors all correlate with eminence in science: early age of first publication, late age of career landmark publications, and age of last publication. Those scientists who are effectively active over the longest time span alter science the most. The crucial adjective is "effectively": It is not publication, but impact that is significant. One thing is certain: Creativity and productivity do not necessarily decline with age.

Some very successful scientists retain a youthful profile of scientific research activity and impact well into old age [26-29].

One other result—not statistically significant, but possibly instructive—is that one of the scientists (no. 32) received a lifetime research award shortly after receiving his doctorate (see Figures). His research was monolithic and was characterized by a single high-impact paper (placing him in impact group 2). Thus, guaranteed funding per se does not appear to spur scientific creativity, and not everyone may benefit from it. Ideas may need money for their development but money does not buy ideas.

References

[1] Szent-Györgyi, A. [Interview]. *The Way of the Scientist.* (New York: Simon & Schuster, 1966) pp. 111-128.

[2] Selye, H., "Biological adaptations to stress." In: Ed. Klemm, W.R., *Discovery processes in modern biology.* (Huntington, NY: Robert Kreiger, 1977) pp. 266-288.

[3] Fehr, H., *Enquête de l'enseignement mathematique [Investigation of mathematical work].* (Paris: Gauthier-Villars, 1912).

[4] Platt, W. & Baker, R.A., "The relationship of the scientific 'hunch' to research."*Journal of Chemical Education*, Vol. 8, 1969 (1931).

[5] Root-Bernstein, R.S., *Discovering.* (Cambridge, MA: Harvard University Press, 1989).

[6]Jewkes, J., Sawers, D., & Stillerman, R., *The sources of invention.* (London: Macmillan, 1958).

[7] Pauling, L., *Linus Pauling: Crusading scientist* [Television interview]. Boston, MA: WGBH-TV. (Transcript of Nova, No. 417) 1977.

[8] Finkelstein, S.N., Scott, J.R., & Franke, A. "Diversity as a contributor to innovative performance by academic physicians." In: (Eds.) Roberts, E.B., Levy, R.I., Finkelstein, S.N., Moskowitz, J. & Sondik, E.J. *Biomedical innovation.* (Cambridge, MA: MIT Press, 1981) pp. 135-143.

[9] Pelz, D.C & Andrews, F.M., *Scientists in organizations: Productive climates for research and development.* (New York: Wiley, 1966) pp. 54-79.

[10] Branscomb, L.M., "The unity of science." *American Scientist*, Vol. 74, 4 (1986).
[11] Gruber, H.E., "Networks of enterprise in creative scientific work." In: Eds. Gholson, B., Houts, A., Neimayer, R.A., & Shadish, W., *Psychology of science and metascience.* (Cambridge, England: Cambridge University Press, 1988).
[12] Gruber, H.E., "The evolving systems approach to creative work." *Creativity Research Journal*, Vol. 1, (1988) 27-51.
[13] Gruber, H.E., *Darwin on man: A psychological study of scientific creativity (2nd ed.).* (Chicago, IL: University of Chicago Press, 1984).
[14] Gruber, H.E., "The evolving systems approach to creative work." In: Eds. Wallace, D.B. & Gruber, H.E., *Creative people at work.* (Oxford: Oxford University Press, 1989) pp. 3 -24.
[15] Diamond, A.M., Jr., "The life-cycle research productivity of mathematicians and scientists." *Journal of Gerontology*, Vol. 41, 520 (1986).
[16] Lehman, H.C., *Age and achievement.* (Princeton, NJ: Princeton University Press, 1953).
[17] Lightman, A.P., "Elapsed expectations." *New York Times Magazine*, March 1984, p. 68.
[18] Simonton, D.K., "Age and outstanding achievement: What do we know after a century of research?" *Psychology Bulletin*, Vol. 104, 251 (1988).
[19] Simonton, D.K., "Career landmarks in science: Individual differences and interdisciplinary contrasts." *Developmental Psychology*, Vol. 27, 119 (1991).
[20] Thomson, G.P., *The strategy of research.* (Surrey: University of Southampton, 1957).
[21] Watson, J.D., [Interview]. *The eighth day of creation.* Judson, H.F. (New York: Simon & Schuster, 1979) pp. 44-45.
[22] Ghiselin, M.T., *Intellectual compromise.* (New York: Paragon House, 1989).
[23]McPherson, J.H., "Prospects for future creativity research in industry." In: Ed. Taylor, C.W., *Widening horizons in creativity.* (New York: Wiley, 1964) pp. 412-424.
[24]Beveridge, W.I.B., *Seeds of discovery.* (New York: Norton, 1980).
[25] Root-Bernstein, R.S. "Elapsed expectations."*New York Times Magazine*, April 1984, p. 130.

[26] Cole, S. "Age and scientific performance." *American Journal of Sociology*, Vol. 84, 958 (1979).
[27] McDowell, J., "Obsolescence of knowledge and career publication profiles: Some evidence of differences among fields in costs of interrupted careers." *American Economic Review*, Vol. 72, 752 (1982).
[28]Stern, N., "Age and achievement in mathematics: A case study in the sociology of science." *Social Studies of Science*, Vol. 8, 127 (1978).
[29]Zuckerman, H., *Scientific elite: Nobel laureates in the United States.* (New York: Basic Books, 1977).

24

Indian Institute of Technology Madras (Chennai) Convocation Speech
Chandrasekhara Venkata Raman

Chandrasekhara Venkata (C.V.) Raman (November 7, 1888 – November 21, 1970)—was an Indian physicist best known for the discovery of the Raman Effect which led to the confirmation of the principles of quantum mechanics in the emission of scattered visible light. He performed this research at the Indian Association for the Cultivation of Science in Calcutta with almost no resources and little modern laboratory equipment. For this he won the Nobel Prize of Physics in 1930. He also did research on the physics of Indian musical instruments. In this speech, he talks to the graduates of the Indian Institute of Technology in Madras (now Chennai) about doing scientific work and the duty of scientists in the modern world and a developing country like India.

Nature and Human Life

I have, in my fairly long life, been at many Convocations, at some of which I received a degree of some sort or the other. I have never seen such an assembly or gathering that so impressed me as this one, which I have been privileged, with your kindness, to address.

Just before I came to the Convocation, the Director was taking me on a 'joy ride' through your campus. I think I should correctly describe this as a joy ride. It was just thrilling, thrilling to see the wonderful old banyan trees or the wild grasses, the thorns here and there and occasionally a 'few' buildings by the way! Well, that

---

* Convocation Address delivered at the Indian Institute of Technology, Madras (Chennai), on 30 July 1966. Reprinted with permission from Aiyasami Jayaraman *C.V. Raman: A Memoir* (New Delhi: Affiliated East-West Press, 1990)

is as it ought to be. Because I always thought that study, examinations, books, lectures and so on are but a very little part of a man's or a woman's, I should add woman's also as I should not forget them, education.

I have always said to myself and others that I regard as the greatest feature of the world Nature herself. She is the supreme artist; she creates forms of beauty, loveliness and colour, unsurpassable, and this has been so from the beginning of time. She is the inspiration not only of artists, painters, sculptors and engineers, but also of men of Science. When I say this, I remember, many years ago, I was standing below the pillars of the temple of Luxor. What did I find at the top? The lotus, papyrus. These forms of beauty of Nature have been the inspiration of all mankind. Well, I should say that they should also be the inspiration of all these graduates of the year.

Usually, technology and industry are associated, I don't say justly, with squalor, dust, ugliness, smoke and all sorts of abomination. That ought not to be so. I think your education is imperfect if you do not realize, my young friends, that life is not merely a question of getting food, clothes and shelter. Man does not live by bread alone. This has been realized from ancient times. I think that the finest things in life are not these, but music, colour, flowers, beauty, aesthetic sense, the satisfaction derived from those. We in Madras do not have to lament about music, I think. You are all music-minded. If you are not, I feel sympathy for you. It is those finer things in life that make life worth living.

You, by the great kindness of no less than three mighty powers, the Central Government of India, the local Government of Madras, to say nothing of that mighty German Republic, have been privileged to live in this wonderful area, with these magnificent hostels, with these great laboratories and, above all, in the midst of these lovely trees and open air. You at least will not die of tuberculosis. I was privileged to be shaken by the hand by some of your prize men. I considered it a privilege to shake hands with them. If I did it with all the graduates, my hands will not be fit to hold anything. They were hefty young fellows with plenty of grip in them. That is as it ought to be. What use are you engineers if you cannot lift up a hammer? Physical strength, energy are the basis of engineering. So, I find, you have been taken care of well.

The German Gift

It is only right and proper that I should make some reference to the great country, Germany, which has helped in no small measure to make it possible for you to receive the education that you have had during these years. To me, when Germany is mentioned, I do not think of Germany on the map. Germany brings to my mind the great masterminds which made Germany what she was. I can mention a score of names. I just mention two who have been recognized as among the greatest of philosophers and men of science the world has ever seen. Hermann von Helmholtz in the nineteenth century and Albert Einstein in the present. I could recite a score of names. Every one of these has written his name in an imperishable way in the records of Science. These, and not the country Germany, nor all that happened to Germany, recall Germany to my mind. It was for many years my earnest desire to go round and spend a few weeks in Germany, visit these ancient centres of learning, small though they may be in geographical extent but great in the lustre of their names. Heidelberg, Gottingen, Marburg and so on. I will never be fortunate enough to spare that time or get that opportunity.

But ten years ago, I received an invitation to attend a conference at Lindau and I thought here was the opportunity for me. I went there, and Lindau is a curious place. How curious it was, I will not mention. It is a place on the edge of the lake we call Lake Constance, the Germans call it Bodensee, whether it is the correct pronunciation, I don't know. Lovely little place and it was a free city of the Empire. And one of the freedoms it possessed was to be allowed to run a casino, or a gambling place. YOU know it is not a very nice kind of freedom. They ran it and made and still make money. People think that when you go to a gambling den you are going to make money. Nothing of the sort. It is the man who keeps the gambling den who makes the money. Lindau made a substantial profit every year and the conscience of the councillors bothered them a little at having this ill-gotten wealth. So what they did was to compromise with conscience. Every year they have a Conference to which they invite nobody other than a Nobel Prize man. A Nobel Prize is the minimum qualification to be invited to this Conference. Year after year, they hold it in succession. Ten years ago, I was invited to this Conference. I went there, I was not sorry I went there, because it is a lovely place. The most beautiful place, right in the middle of the lake, is an island called the island of Mainau. Mainau is owned by Count Bernadotte. He comes from the Swedish Royal family. He was the host of this function.

Afterwards I moved on to that old university, one of the very first universities which gave an honorary degree, the ancient University of Freiburg in Breisgau. I was there for a week and then I moved on to Bonn. And then from Bonn to Munich before I came out of Germany. I am mentioning this because I was in Bonn for just a week. I had a wonderful time. You know, at that time, Germany had, even in 1956, not recovered from the devastations of the War. They were working very hard to clean up the devastation. I was enormously impressed by the Museum of Mineralogy which had been set up in Bonn. It is absolutely incredible how from a devastated country they could get together such amazing beautiful collections of rare and beautiful specimens. That is one of my happy experiences. Others I will not mention.

And on that occasion, it so happened—all this is introductory to my remarks that when I was there, our late Prime Minister, Jawaharlal Nehru, was also there. I want you to realize that he was not there because I was there. Nor was I there because he was there. A pure accidental coincidence. Such coincidences always happen, as you know. I think the Indian Ambassador there thought that this opportunity should not be missed. So I was invited to a lunch in the President's house at which the President of the German Republic, Mr. Nehru, the Indian Ambassador and a few others sat round the table. One of the curious things that happened at that meeting was the President made a speech in German which lasted about 5 to 10 minutes. And then, at the end of it, the interpreter got up and translated word for word the whole of the 10 minute speech in English. Then Mr. Nehru got up and made a speech in English, The interpreter got up and translated the whole speech word for word into German. But I must confess that I do not remember at this distance of time what exactly they spoke. But I presume it was the usual declarations of mutual love and affection which are always made when great dignitaries meet. Why I am mentioning this is because, it was on that occasion, I read from your book, that the German Republic promised to give a gift of this Institute.

Well, just ten years ago, imagine ten years ago. I find it very difficult to, understand how in the course of ten years a veritable jungle full of, presumably, snakes and cobras has been transformed into this beautiful place of learning and how so many buildings and so much equipment and so many bright young people from all over India have gathered together and such a magnificent assembly presented for my delectation. I enjoy this, particularly because I love colour. As I have mentioned on many occasions, there is an unwritten law that men should not wear bright colours as a rule. Women are allowed to wear colours as you know and they always do. But there are exceptions. And a convocation is one of these exceptions when the men are allowed to flaunt all the bright colours in the spectrum and a few outside the spectrum as well, for the delectation and admiration of those round them and specially members of the feminine sex who will be around.

Youth and Freshness of Outlook

Well, as I say, it is a very enjoyable occasion and I think my young friends are all extremely happy to have had this great occasion. It must stand out in their memory for a long time to come as the unique occasion in their life. Well, I remember myself, you see my mind goes back to a period sixty years ago, fully sixty years ago, I also came out of a college which is in the same town, with a degree in my pocket arid a few other things, I won't mention what. Now that occasion stands out for one reason and I want to mention that reason. Looking back over the years, I find to my astonishment, to my surprise, that the experiences which I went through in those four years have left an indelible impression on my mind, an impression that sixty years of time and all that has happened since has not succeeded in dimming in the least. What is even more remarkable is this—it is perfectly true to say that what I am today, what I have done in the last sixty years, has all been determined for me with absolute mathematical precision by what I did in those four years. The opportunities that I had during those years, made me turn my mind to certain things. I have found it impossible to turn away from these, because of the force in me of those youthful days. I was only 14 years of age when I entered the Presidency College and 18 when I came out with my Master's degree and an appointment as an Officer of the Finance Department, and a published paper as a budding scientist. All these at the age of 18! At that early age, the mind is so impressionable. And what I did then had determined my whole career.

I want to stress this because I want you, my young friends, to realize that in these years, three or four or five years as the case may be, you have been subjected to the influence of a band of teachers, you have been subjected to the influence of the old banyan trees around, which I don't regard as unimportant. Now these, please do remember that, in fact, are going to determine your future career. But that is not all. What you are going to be, my young friends, depends upon what you are going to do in the next few years to come. Alas! It is often very true that people get a degree, get a job and perhaps get married, and then forget all about what they learned in college. That is hardly the thing to do. If you are all going to be worth any little at all in the future, it can be only if you remember what has been laid now as the foundation in these four or five years. On that foundation you must build.

You always have to remember a few things, you will permit me to remind you of them. This most wonderful possession that you all have, everyone of us has, is this human body. It is our parents who gave it to us. I have recently turned my attention from physics and chemistry, mineralogy and mathematics to the study of the human faculties. Some of them, unfortunately, are hidden away inside the brain and we have to take them for granted. The mere study of all these external points of contact has made me realize what an amazing possession we have.

There is another little thing that we all forget very often. There is a little thing called the 'heart.'. That little machine started working when you were born, why even before you were born and goes on ticking at a certain rate and it goes on ticking, ticking, ticking all the time at the same rate, at nearly the same rate, all the time when you are young and all the time when you are getting to be an old man, until you are dead. And when that stops, you are dead. That wonderful machine, my dear young friends, has to be safeguarded. You often hear of great men suddenly dropping down, their doctors call it coronary thrombosis or whatever other names, learned names for this collapse which causes death. And why is it that this happens? It is because they over-drive this wonderful machine. They misuse their bodies. They think, now that I am so young and energetic, I can do anything I please, I can eat all the chillies I want, go out all night, attend theatres, friendship parties and so on. And what happens? No doubt the young blood can stand the strain, but it tells. And what happens then? You are prematurely aged and then coronary thrombosis comes in and takes you away. I want you to realize this at the time of youth. This is a lesson I have learned myself. Please do not think that I am preaching what I don't practise. I have always practised it. I believe not in precept but in practice. Always. I believe that the greatest influence that a teacher has is, as they say, by his example and not by his perception only. Now is the time when you are still young, when the blood is still flowing warmly is your veins and arteries, now is the time to improve on what you learned here.

India has an enormous population. We are all trying to make India great. But who is going to make it great? It is only the young intelligentsia of the country. Realize it is up to them to use not only their hands but use their brains to learn to think. The faculty of independent thinking must be applied to all problems of life. YOU must be serious about life. You must not think that hasty pleasures or indulgence in all sorts of loose things are going to help you at all. This is the lesson that I have learned in these sixty years of life, actively as a man of Science. There is nothing that makes a man so happy as some real achievement. It is the achievement of doing something real that has a permanent value and will be recognized all the world over. The money bags that you find in the Reserve Bank are nothing when compared to this achievement. The mere joy of achievement is something very great. And I think all our young people who come out of our Universities realize that it is up to them to see how best they can raise the glory of India and how best they can make themselves happy, how best they can achieve even material success. It is only by realizing this while they are young, while they have energy, this will be possible.

I am almost inclined to enter into a dissertation on old age versus youth. I could speak for half an hour on that thesis, 'Age versus Youth.' You know that age is credited with wisdom. With all deference to the people around, I beg to question that proposition. I tell you what youth brings along with it. If you do not take care of yourself, you are left with creaky bones, your teeth drop out, your eyes become blurred, your ears become half deaf and, worst of all, you become cynical, contemptuous of others. You come to think life is not worth living. And, in fact, you come almost to such a stage that but for the unfortunate desire we all have to continue to live, the easiest plan would be to swallow a tube of morphia and be done with it all. We do not feel like it and that is because God has implanted in us an absolutely unhealthy desire to continue life in spite of these miseries. I tell you that because it is very, very difficult indeed to summon up, as you grow old, those enthusiasms, those fears, desire for achievement, energy and all that. Youth is the most glorious time of all. I have said elsewhere that most of the great discoveries in Science have been made by young people. It is not the experience or wisdom that old age brings, but the freshness of outlook, the indomitable desire to achieve, which is the characteristic of youth, that makes discoveries possible. It is this that makes life worthwhile. If only you realize this and realize that here I am, I am still young, let me see what I can do, that all discoveries become possible.

Fearless and Independent Thinking

And above all, in my own experience, I have found that one of our evils is that for centuries we have been trodden under the feet of conquerors from abroad. I don't want to recite all their names. One of the things that has been bred in us, a very deep and ineradicable defect, is a kind of inferiority complex which makes us think that we dare not question what has come to us from abroad. Whatever comes to us in a textbook must be right, some great man has said something, well, we must bow to him in fear and trembling and never question him. This produces a mental inhibition. Now, I do not suggest to you that you should all become arrogant, contemptuous of all the great men of the past. Not at all. I am not suggesting that. But I think we should all learn that no one is infallible. Not even Hermann von Helmholtz, not even Einstein. Nobody is infallible. New knowledge may upset what may have been made in the past, and may completely throw out what has been done before.

So I think one of the things all Indians should learn is fearless independence of thinking. That is a quality which is very essential and the absence of which, if I may venture to say so with all deference, is what stands today in the path of Indian progress. Whenever we want to do anything, we borrow money from abroad. We all know that everyday we see one hundred million dollars or one billion dollars being borrowed from somewhere. Our so-called independence freely consists, if I may say so without entering into politics, in our being ruled by a consortium of all the nations in the world, except ourselves. Leave that alone. Bad enough to borrow money. But what about borrowing knowledge? What about borrowing experts from abroad? What about forgetting to think for ourselves? This feeling of helplessness must be shaken away, shaken out ruthlessly. We must realize that we must stand on our own legs. It is better to work with the most inefficient useless equipment of ours than to shine in borrowed feathers, better to work on problems with our slender resources. We must realize this and until and unless we realize this, we cannot go on. And I want to turn to my great industrialist friends who want to buy know-how at great expense from abroad. Let me assure them that they will never get on and get very far. You see that the signs are already there. The rupee has come down, I do not know to how many cents, and Mr. Masani has said that it will soon come down to five cents. I am not wishing that his prophecy comes true.

Progress of Science and Technology

But no country, and as you are all engineers, let me express here forcibly my conviction that no country can become industrially great without a foundation of real knowledge. This is what Science teaches. Science has shown time and again that Science comes first and Technology afterwards. Without Science there is no Technology. Why has Germany been so great? Because in the nineteenth century, she had a galaxy of men of Science in every branch of knowledge, whose name and fame shone forth. Because they were not technicians, they were humble professors in the universities. But they sought knowledge and they made their students seek knowledge. They were springs from which knowledge came forth, gushed forth and it was that knowledge, that spring of knowledge, that fertilized all the industries of Germany and made her great.

It is realized in all countries today, that Science comes first and Technology afterwards. If you think you can build a great industrial nation, make tons of money and pay off all these awful debts by pursuing so-called Technology alone, you are doomed to complete failure. Let me say this without hesitation. It is only when we set our houses in order and build up powerful schools of thinking in every field, electricity, chemistry, metallurgy and so on, only then we will have the solid basis of knowledge from which can come forth men who will teach your technologists what to do. I think it is true to say that the finest instruments, the most sophisticated instruments, are not found in technological laboratories, but are found in the research laboratories where men are trying to explore the unknown world and try to discover things. There is also another thing which is well worth remarking upon and that is, in not a few cases, not only is Science the fountainhead of technological knowledge, in many cases Science has set the problems which the technologists had to solve.

You may recall the history of Astronomy of the nineteenth century. Astronomy is usually regarded as one of the useless subjects. Our ancients were much concerned about the stars because they thought that the sun and the stars had something to do with human affairs. So they very keenly watched the planets and the stars. Though the reasons why they may have done so may be wrong, there is no doubt that the pursuit of astronomy is of infinite importance to us. The study of astronomy may look as if it has nothing to do with human affairs, if you do not believe in astrology, but really it is more from the study of the stars that you have learned more about the earth we live in, more about the sun and more about everything, than from the study of terrestrial sources of light. The astronomers demanded instruments of the highest precision with which alone they could follow the movement of the stars, do the time-keeping, note the displacements, aberrations, parallax and so on. The demands of precision instruments, especially by Germans like Bessel and others, led to the modern development of precision to mention the fact that the great telescope at Palomar, the 200-inch telescope, that enormous thing, that tremendous, huge, massive thing, hundreds of tons in weight, has to be moved with the accuracy and precision of a Swiss watch and nothing less. Nothing less will satisfy the astronomers. This huge mass has to go round with the smoothness and precision of a Swiss watch. I know what the precision of a Swiss watch is. I have got here a watch on my wrist which I wear only on occasions such as this. Well, I want to be reminded of the time which passes and I wind it only about once a year or so. But all the time it shows the correct time. Astronomy demands such precision and it was the demand of the astronomers that led to the development of precision mechanics which, of course, today we use in everything.

You can quote example after example to prove this. The demands of Science, the botanist, the zoologist who wanted to examine his subtle structures led to the development of the great firm of Carl Zeiss. It was Ernest Abbe who took up that problem and made the optical industry what it is today. What was demanded and made for the needs of Science has benefited everybody else. The same story is repeated in all the beautiful instruments that have gone to make this great advancement of knowledge possible in these sixty years. They were born out of the brain of men of Science, they were translated into practice and today they are the tools of study in every branch of knowledge. Every metallurgist today uses the electron microscope. Who thought of it? A man of Science who was not interested in its possible applications. He was interested in it for its own sake.

Now I wind up. I think I have gone on long enough. I must stop. I know you are all listening to me with great eagerness and interest. I must put an end to your agonies. I will do so as soon as I can. But this I would like to say, that the development of knowledge and Science in the last sixty years, I have seen from the inside. I have not been inside all the time, but I have been inside most of the time. It has been something absolutely fantastic.

The Nature of Scientific Advances

Why is it that Science has developed in such an explosive or spectacular fashion in the last sixty years? There are really three causes. We can analyse them and put them apart. First and foremost stands the fact that towards the end of the nineteenth century, or in the beginning of this century, came a succession of epochmaking discoveries in fundamental knowledge, the discovery of the quantum of action by Planck, the discovery of Einstein of the corpuscular nature of light and then the use of this principle by Niels Bohr and others to analyse the structure of the atom, to explore the chemical molecule. This sort of explosive development at the beginning of the century made a terrific advance possible, in the study of the ultimate structure of matter, and this explosion is still going on.

There is a second and a great stream of scientific knowledge which has come out in this way. And that is the fact that Science, as we all know and recognize it, has very definite applications to human welfare. This is most evident from examples in the field of agriculture, the knowledge of heredity, the laws of Mendelism and so on. They made a great difference to the science of agriculture. So also, one of the most important advances which has a bearing on the work of you gentlemen is the discovery of what is known as plastics. The whole science of micro-electro chemistry as it is called has developed from the work of a man who was interested simply in studying the big molecules, Staudinger. Then the subject took on an explosive development. Today we can't live without plastics. We drink our coffee out of plastic cups. We wear artificial silks made of nylon and plastics. It is a typical example of how pure Science has led to the development of vast industries with the greatest possible importance. Then we have also examples in medicine. From ancient times, the human body has been subjected to all kinds of illnesses and the 'medicine men' with their drugs did a roaring practice. The modern expression with regard to those funny people is 'witch doctors'. From the medicines of these witch doctors to the modern medicines, the leap has been marvellous.

But there is a third and a most sinister way in which Science has developed in the last sixty years. And that is Science in its applications to war, application to defensive war and offensive war. Even in medicine, much of our knowledge of the human brain has come from the study of half-dead and half-dying men in war. The examination of their bodies has led to a profound advance in medical knowledge. So war has had that result. I think we can say that a lot of modern Science has really come directly out of the needs of war. I knew for example, in the First World War, men like Lord Rutherford, Sir William Bragg and many other British scientists were all busy trying to combat the submarine menace by the invention of the subsonic devices that today are being used everywhere. Even in the First World War, the use of aviation in warfare was already being developed; in the Second World War it took a tremendous turn.

And not only that, the Second World War, as you all know, brought the atom bomb into existence. That atom bomb has come directly out of the consequence of the actual discovery of fission made by Joliot Curie in the laboratory—he showed it to me when I was there in Paris, the gist of the discoveries made. That immediately set thinking minds in action here is an instrument of dreadful power, if we use it we can destroy mankind. And fear, the fear that the other man may use it led to the development of the atom bomb everywhere. Then the hydrogen bomb came. And ever since, an atmosphere of fear, it is a horrible thing to see, an atmosphere of fear of mutual recrimination, progressive deterioration is there, like what happens to a man when he borrows money. You see the interest goes on adding up. It becomes a colossal figure, which bears no resemblance whatsoever to the loan which he took at first. This kind of explosive development of fear complex has produced a psychological, a pathological state of affairs in the human mind in which all evils thrive and sustain. Today, Science in many countries is simply the handmaid of the war machine.

Wonderful achievements, rockets sent up and two men jumped out of it and had a rope tied to their bellies and they walked in space. And everybody says: 'Oh! What a great feat!' Let me tell you, I simply smile with loathing and contempt. It is with feelings of loathing and contempt with which I witness this colossal display of lunacy on the part of mankind. It is nothing less than lunacy, sheer raving lunacy, to spend billions of dollars. Instead of shooting two men into space they could have shot two monkeys and made them walk in space. It is just a mere pretence. I say this with all sense of responsibility, it is a mere pretence to say that all these exploits of finding out what exactly is there on the moon and so on, have any scientific value whatever. I absolutely deny that. It is nothing but militarism very thinly disguised. That is what is happening today. It is very sad. Our Science is going that way. So Science today is misused.

I heard the pledge given by you, that you will not use your knowledge for unworthy ends. If you had been in one of those countries you would have to use it for unworthy ends. Otherwise you would lose your job forthwith. What is the use of giving a pledge which you would have to break in order to get your daily bread? This is what is happening today. There are sensitive consciences even in those countries who revolt from this sort of thing. But it is going on, going on, this prostitution of Science. Where it will end, I do not know.

Only I want to say this, that we in India to some extent are slaves. We are part of the machinations of our so-called friends. We are forced to accept this situation. Our friends in the United States give all the latest arms to our friends in Pakistan and they want them to use these against us and they have used these against us and we are forced to reply in time. The Russians thought that the Chinese were a friendly nation. They taught the Chinese all the arts of war. What China is doing today is what Russia had taught her to do, and now China is paying her back in her own coin. This is a horrible situation. I do not know what to do. As a man of Science, my heart is simply wrung with this amazing prostitution of Science. We can do nothing about it. In fact, we have to accept the situation as it is and do what we can. I am sorry to have had to end on this very unfortunate and depressing note. There it is.

Self-Reliance and the Need of the Hour

But let me end by saying that we in this country have no future whatever of any sort unless we learn firstly, secondly and lastly to rely on ourselves for everything that we need. It is better, I think, to go back to the Gandhian age and ride an ox-cart, to throw away radio, television and everything and go back to the land of the ox-cart. We cannot do it unfortunately. We are tied to the coat-tails of European civilization, I include myself also in this. In the first place I put a big query mark after the word 'civilization' and I must also put another mark after 'European'. I must also add American, you know the American way of civilization. We are tied to their coat-tails. We find that we cannot be happy unless we have a radio making a lot of noise in the other room. I never listen to a radio. I simply loathe it. One of the things that we have been taught from childhood is to admire this wonderful flicker on the screen. I never go to a cinema, never. For twenty years I have not stepped inside a cinema theatre. I cannot advise my young friends to follow my example because I know they won't follow my example. We are told that we cannot be happy unless we have a television set to see some lady dancing on the television screen. This is the trouble,

You see our tastes have been corrupted. We do not want to look at the banyan trees. We must go and see the cinema screen. Now we must, I say this with all energy that I can command, we have to eschew all the evil things which we learn from Europe and America. Let us not condemn Science for that reason, but we must not subordinate ourselves to the ideas and ideals which have come to us from the West, which are simply designed to make us part with our rupees and make the rupee worth five cents, as Mr. Masani says.

If we cannot do these things ourselves, let us do without them, as I say, let us walk, let us go in the country carts. If we cannot make our motor cars, why should we buy them from abroad? Why even import parts? It is one of those funny things, when it is said, "Oh! 85 per cent is 'local made' and the remaining 15 per cent is 'imported'." They talk of electronics. I have tried hard to find out if there is any one place—I have not yet heard an answer to that—if there is any place in India where the most important component of all electronic valves is made (Ed. Note: by 'electronic valves' Raman is probably referring to p-n junction devices such as diodes and transistors). All these depend upon the manufacture of a metallic filament which will stand the current. If they are being made in India I should like to know where they have been made. I have not heard of any. That is the starting point of the whole electronic industry and do we make it? Let us wait till we can make it, before we buy a single electronic valve from outside. If we do that, then I think we should learn how to make it. It is this lesson of self-reliance that we have to learn and until we learn it there is no future for us.

Part V
Ethics & Cranks

25

Ten Simple Rules for Building and Maintaining a Scientific Reputation

Philip E. Bourne & Virginia Barbour

Virginia Barbour is the chief editor for PLoS Medicine and Chair of the Committee on Publication Ethics (COPE). She began her research in hematology, later becoming an editor for the journal The Lancet and is one of the founders of PLoS Medicine.

While we cannot articulate exactly what defines the less quantitative side of a scientific reputation, we might be able to seed a discussion. We invite you to crowd source a better description and path to achieving such a reputation by using the comments feature associated with this article. Consider yourself challenged to contribute.

At a recent Public Library of Science (PLoS) journal editors' meeting, we were having a discussion about the work of the Committee on Publication Ethics (COPE; http://www.publicationethics.org/), a forum for editors to discuss research and publication misconduct. Part of the discussion centered on the impact such cases have on the scientific reputation of those involved. We began musing: What on earth is a scientific reputation anyway? Not coming up with a satisfactory answer, we turned to a source of endless brainpower—students and other editors. Having posed the question to a group of graduate students, PLoS, and other editors, we got almost as many different answers

* Bourne, P.E. & Barbour, V., "Ten Simple Rules for Building and Maintaining a Scientific Reputation." *PLoS Comput Biol* Vol. 7, e1002108 (2011).

as people asked, albeit with some common themes. They all mentioned the explicit elements of a reputation that relate to measurables such as number of publications, H factor (h-index), overall number of citations etc., but they also alluded to a variety of different, qualitative, factors that somehow add up to the overall sense of reputation that one scientist has for another.

What these students and editors identified en masse is one important side of a scientific reputation that is defined by data; but they also identified a much more nebulous side, that, while ill-defined, is a vital element to nurture during one's career. A side defined to include such terms as fair play, integrity, honesty, and caring. It is building and maintaining this kind of less tangible reputation that forms the basis for these Ten Simple Rules. You might be wondering, how can you define rules for developing and maintaining something you cannot well describe in the first place? We do not have a good answer, but we would say a reputation plays on that human characteristic of not appreciating the value of something until you do not have it any more.

A scientific reputation is not immediate, it is acquired over a lifetime and is akin to compound interest—the more you have the more you can acquire. It is also very easy to lose, and once gone, nearly impossible to recover. Why is this so? The scientific grapevine is extensive and constantly in use. Happenings go viral on social networks now, but science has had a professional and social network for centuries; a network of people who meet each other fairly regularly and, like everyone else, like to gossip. So whether it is a relatively new medium or a centuries-old medium, good and bad happenings travel quickly to a broad audience. Given this pervasiveness, here are some rules, some intuitive, for how to build and maintain a scientific reputation.

Rule 1: Think Before You Act

Science is full of occasions whereupon you get upset—a perceived poor review of a paper, a criticism of your work during a seminar, etc. It is so easy to immediately respond in a dismissive or impolite way, particularly in e-mail or some other impersonal online medium. Don't. Think it through, sleep on it, and get back to the offending party (but not a broader audience as it is so easy to do nowadays with, for example, an e-mail cc) the next day with a professional and thoughtful response, whatever the circumstances. In other words, always take the high road whatever the temptation. It will pay off over time, particularly in an era when every word you commit to a digital form is instantly conveyed, permanently archived somewhere, and can be retrieved at any time.

Rule 2: Do Not Ignore Criticism

Whether in your eyes, criticism is deserved or not, do not ignore it, but respond with the knowledge of Rule 1. Failure to respond to criticism is perceived either as an acknowledgement of that criticism or as a lack of respect for the critic. Neither is good.

Rule 3: Do Not Ignore People

It is all too easy to respond to people in a way that is proportional to their perceived value to you. Students in particular can be subject to poor treatment. One day a number of those students will likely have some influence over your career. Think about that when responding (or not responding). As hard as it is, try to personally respond to mail and telephone calls from students and others, whether it is a question about your work or a request for a job. Even if for no other reason, you give that person a sense of worth just by responding. Ignoring people can take other serious forms, for example in leaving deserving people off as paper authors. Whether perceived or real, this can appear that you are trying to raise your contribution to the paper at the expense of others—definitely not good for your reputation.

Rule 4: Diligently Check Everything You Publish and Take Publishing Seriously

Science does not progress in certainties—that is one of its joys but also what makes it such a hard profession. Though you cannot guarantee that everything you publish will, in 50 years' time, be shown to be correct, you can ensure that you did the work to the accepted standards of the time and that, whether you were the most junior or senior author, you diligently checked it (and checked it again…) before you submitted it for publication. As a first author you may well be the only one who appreciates the accuracy of the work being undertaken, but all authors have a responsibility for the paper. So, however small or big your contribution, always be upfront with your co-authors as to the quality and accuracy of the data you have generated. When you come to be a senior author, it is so easy to take a draft manuscript at face value and madly publish it and move on. Both actions can come back to haunt you and lead to a perception of sloppy work, or worse, deception. As first author, this mainly lets down your other authors and has a subtle impact on your growing reputation. As the senior author of an error-prone study, it can have a more direct and long-lasting impact on your reputation. In short, take publication seriously. Never accept or give undeserved authorship and in addition never leave anyone out who should be an author, however lowly. Authorship is not a gift—it must be earned and being a guest or gift author trivializes the importance of authorship. Never agree to be an author on a ghostwritten paper. At best these papers have undeclared conflicts of interest; at worst potential malpractice.

Rule 5: Always Declare Conflicts of Interest

Everyone has conflicts of interest, whether they are financial, professional, or personal. It is impossible for anyone to judge for himself or herself how their own conflict will be perceived. Problems occur when conflicts are hidden or mismanaged. Thus, when embarking on a new scientific endeavor, ranging from such tasks as being a grant reviewer, or a member of a scientific advisory board, or a reviewer of a paper, carefully evaluate what others will perceive you will gain from the process. Imagine how your actions would be perceived if read on the front page of a daily newspaper. For example, we often agree to review a paper because we imagine we will learn from the experience. That is fine. Where it crosses the line is when it could be perceived by someone that you are competing with the person whose work you are reviewing and have more to gain than just general knowledge from reviewing the work. There is a gray area here of course, so better to turn down a review if not sure. Failure to properly handle conflicts will eventually impact your reputation.

Rule 6: Do Your Share for the Community

There is often unspoken criticism of scientists who appear to take more than they give back. For example, those who rarely review papers, but are always the first to ask when the review of their paper will be complete; scientists who are avid users of public data, but are very slow to put their own data into the public domain; scientists who attend meetings, but refuse to get involved in organizing them; and so on. Eventually people notice and your reputation is negatively impacted.

Rule 7: Do Not Commit to Tasks You Cannot Complete

It tends to be the same scientists over and over who fail to deliver in a timely way. Over an extended period, this becomes widely known and can be perceived negatively. It is human nature for high achievers to take on too much, but for the sake of your reputation learn how to say no.

Rule 8: Do Not Write Poor Reviews of Grants and Papers

Who is a good reviewer or editor is more than just perception. Be polite, timely, constructive, and considerate and, ideally, sign your review. But also be honest—the most valued reviewers are those who are not afraid to provide honest feedback, even to the most established authors. Editors of journals rapidly develop a sense of who does a good job and who does not. Likewise for program officers and grant reviews. Such perceptions will impact your reputation in subtle ways. The short term gain may be fewer papers or grants sent to you to review, but in the longer term, being a trusted reviewer will reflect your perceived knowledge of the field. Although the impact of a review is small relative to writing a good paper in the field yourself, it all adds up towards your overall reputation.

Rule 9: Do Not Write References for People Who Do Not Deserve It

It is difficult to turn down writing a reference for someone who asks for one, even if you are not inclined to be their advocate; yet, this is what you should do. The alternative is to write a reference that (a) does not put them in a good light, or (b) over exalts their virtues. The former will lead to resentment; the latter can impact your reputation, as once this person is hired and comes up short, the hirer may question aspects of your own abilities or motives.

Rule 10: Never Plagiarize or Doctor Your Data

This goes without saying, yet it needs to be said because it happens, and it is happening more frequently. The electronic age has given us tools for handling data, images, and words that were unimaginable even 20 years ago, and students and postdocs are especially adept in using these tools. However, the fundamental principle of the integrity of data, images, and text remains the same as it was 100 years ago. If you fiddle with any of these elements beyond what is explicitly stated as acceptable (many journals have guidelines for images, for example), you will be guilty of data manipulation, image manipulation, or plagiarism, respectively. And what is more, you will likely be found out. The tools for finding all these unacceptable practices are now sophisticated and are being applied widely. Sometimes the changes were done in good faith, for example, the idea of changing the contrast on a digital image to highlight your point, but one always needs to think how such a change will be perceived and in fact whether it might, even worse, give the average reader a false sense of the quality of that data. Unfortunately, even if done in good faith, if any of these practices are found out, or even raised as a suspicion, the impact on one's career can be catastrophic.

In summary, there are a number of dos and don'ts for establishing a good reputation—whatever that might be. Do not hesitate in giving us your thoughts on what it means to be a reputable scientist.

26

Cargo Cult Science
Richard Feynman

Richard Feynman (May 11, 1918 – February 15, 1988)—was an American physicist who is known for his work in quantum electrodynamics and the path integral formulation of quantum mechanics. Recognized as truly brilliant by J. Robert Oppenheimer, he was brought onto the Manhattan Project and helped build the first atomic bomb. Even then his quirky and mischievous nature was evident where he would crack safes and circumvent base security as practical jokes. He eventually became a professor at Caltech and his work eventually won him the Nobel Prize in 1965 with Julian Schwinger and Sin-Itiro Tomonaga. He is popularly well-known as an archetype for a scientific genius and has been popularized in a series of books, most famously a biography by James Gleick, and his recorded lectures on physics The Feynman Lectures. In probably his most famous writing, he talks about 'cargo cult science' which appears like real science but lacks the essence of true scientific investigation.

During the Middle Ages there were all kinds of crazy ideas, such as that a piece of rhinoceros horn would increase potency. Then a method was discovered for separating the ideas—which was to try one to see if it worked, and if it didn't work, to eliminate it.

This method became organized, of course, into science. And it developed very well, so that we are now in the scientific age. It is such a scientific age, in fact that we have difficulty in understanding how witch doctors could *ever* have existed, when nothing that they proposed ever really worked—or very little of it did.

But even today I meet lots of people who sooner or later get me into a conversation about UFOs, or astrology, or some form of mysticism, expanded consciousness, new types of awareness,

---

* Adapted from the Caltech commencement address given in 1974.

ESP, and so forth. And I've concluded that it's *not* a scientific world.

Most people believe so many wonderful things that I decided to investigate why they did. And what has been referred to as my curiosity for investigation has landed me in a difficulty where I found so much junk that I'm overwhelmed. First I started out by investigating various ideas of mysticism, and mystic experiences. I went into isolation tanks and got many hours of hallucinations, so I know something about that. Then I went to Esalen, which is a hotbed of this kind of thought (it's a wonderful place; you should go visit there). Then I became overwhelmed. I didn't realize how *much* there was.

At Esalen there are some large baths fed by hot springs situated on a ledge about thirty feet above the ocean. One of my most pleasurable experiences has been to sit in one of those baths and watch the waves crashing onto the rocky shore below, to gaze into the clear blue sky above, and to study a beautiful nude as she quietly appears and settles into the bath with me.

One time I sat down in a bath where there was a beautiful girl sitting with a guy who didn't seem to know her. Right away I began thinking, "Gee! How am I gonna get started talking to this beautiful nude babe?" I'm trying to figure out what to say, when the guy says to her, "I'm, uh, studying massage. Could I practice on you?" "Sure," she says. They get out of the bath and she lies down on a massage table nearby.

I think to myself, "What a nifty line! I can never think of anything like that!" He starts to rub her big toe. "I think I feel it, "he says. "I feel a kind of dent—is that the pituitary?"

I blurt out, "You're a helluva long way from the pituitary, man!"

They looked at me, horrified—I had blown my cover—and said, "It's reflexology!"

I quickly closed my eyes and appeared to be meditating.

That's just an example of the kind of things that overwhelm me. I also looked into extrasensory perception and PSI phenomena, and the latest craze there was Uri Geller, a man who is supposed to be able to bend keys by rubbing them with his finger. So I went to his hotel room, on his invitation, to see a demonstration of both mindreading and bending keys. He didn't do any mindreading that succeeded; nobody can read my mind, I guess. And my boy held a key and Geller rubbed it, and nothing happened. Then he told us it works better under water, and so you can picture all of us standing in the bathroom with the water turned on and the key under it, and him rubbing the key with his finger. Nothing happened. So I was unable to investigate that phenomenon.

But then I began to think, what else is there that we believe? (And I thought then about the witch doctors, and how easy it would have been to cheek on them by noticing that nothing really worked.) So I found things that even *more* people believe, such as that we have some knowledge of how to educate. There are big schools of reading methods and mathematics methods, and so forth, but if you notice, you'll see the reading scores keep going down—or hardly going up in spite of the fact that we continually use these same people to improve the methods. *There's* a witch doctor remedy that doesn't work. It ought to be looked into; how do they know that their method should work? Another example is how to treat criminals. We obviously have made no progress—lots of theory, but no progress—in decreasing the amount of crime by the method that we use to handle criminals.

Yet these things are said to be scientific. We study them. And I think ordinary people with commonsense ideas are intimidated by this pseudoscience. A teacher who has some good idea of how to teach her children to read is forced by the school system to do it some other way—or is even fooled by the school system into thinking that her method is not necessarily a good one. Or a parent of bad boys, after disciplining them in one way or another, feels guilty for the rest of her life because she didn't do "the right thing," according to the experts.

So we really ought to look into theories that don't work, and science that isn't science.

I think the educational and psychological studies I mentioned are examples of what I would like to call cargo cult science. In the South Seas there is a cargo cult of people. During the war they saw airplanes land with lots of good materials, and they want the same thing to happen now. So they've arranged to imitate things like runways, to put fires along the sides of the runways, to make a wooden hut for a man to sit in, with two wooden pieces on his head like headphones and bars of bamboo sticking out like antennas—he's the controller—and they wait for the airplanes to land. They're doing everything right. The form is perfect. It looks exactly the way it looked before. But it doesn't work. No airplanes land. So I call these things cargo cult science, because they follow all the apparent precepts and forms of scientific investigation, but they're missing something essential, because the planes don't land.

Now it behooves me, of course, to tell you what they're missing. But it would be just about as difficult to explain to the South Sea Islanders how they have to arrange things so that they get some wealth in their system. It is not something simple like telling them how to improve the shapes of the earphones. But there is *one* feature I notice that is generally missing in cargo cult science. That is the idea that we all hope you have learned in studying science in school—we never explicitly say what this *is*, but just hope that you catch on by all the examples of scientific investigation. It is interesting, therefore, to bring it out now and speak of it explicitly. It's a kind of scientific integrity, a principle of scientific thought that corresponds to a kind of utter honesty—a kind of leaning over backwards. For example, if you're doing an experiment, you should report everything that you think might make it invalid—not only what you think is right about it: other causes that could possibly explain your results; and things you thought of that you've eliminated by some other experiment, and how they worked—to make sure the other fellow can tell they have been eliminated.

Details that could throw doubt on your interpretation must be given, if you know them. You must do the best you can—if you know anything at all wrong, or possibly wrong—to explain it. If you make a theory, for example, and advertise it, or put it out, then you must also put down all the facts that disagree with it, as well as those that agree with it. There is also a more subtle problem. When you have put a lot of ideas together to make an elaborate theory, you want to make sure, when explaining what it fits, that those things it fits are not just the things that gave you the idea for the theory; but that the finished theory makes something else come out right, in addition.

In summary, the idea is to try to give *all* of the information to help others to judge the value of your contribution; not just the information that leads to judgment in one particular direction or another.

The easiest way to explain this idea is to contrast it, for example, with advertising. Last night I heard that Wesson oil doesn't soak through food. Well, that's true. It's not dishonest; but the thing I'm talking about is not just a matter of not being dishonest, it's a matter of scientific integrity, which is another level. The fact that should be added to that advertising statement is that no *oils* soak through food, if operated at a certain temperature. If operated at another temperature, they *all* will—including Wesson oil. So it's the implication which has been conveyed, not the fact, which is true, and the difference is what we have to deal with.

We've learned from experience that the truth will come out. Other experimenters will repeat your experiment and find out whether you were wrong or right. Nature's phenomena will agree or they'll disagree with your theory. And, although you may gain some temporary fame and excitement, you will not gain a good reputation as a scientist if you haven't tried to be very careful in this kind of work. And it's this type of integrity, this kind of care not to fool yourself, that is missing to a large extent in much of the research in cargo cult science.

A great deal of their difficulty is, of course, the difficulty of the subject and the inapplicability of the scientific method to the subject. Nevertheless it should be remarked that this is not the only difficulty. That's why the planes *don't* land—but they don't land.

We have learned a lot from experience about how to handle some of the ways we fool ourselves. One example: Millikan measured the charge on an electron by an experiment with falling oil drops, and got an answer which we now know not to be quite right. It's a little bit off, because he had the incorrect value for the viscosity of air. It's interesting to look at the history of measurements of the charge of the electron, after Millikan. If you plot them as a function of time, you find that one is a little bigger than Millikan's, and the next one's a little bit bigger than that, and the next one's a little bit bigger than that, until finally they settle down to a number which is higher.

Why didn't they discover that the new number was higher right away? It's a thing that scientists are ashamed of—this history—because it's apparent that people did things like this: When they got a number that was too high above Millikan's, they thought something must be wrong—and they would look for and find a reason why something might be wrong. When they got a number closer to Millikan's value they didn't look so hard. And so they eliminated the numbers that were too far off, and did other things like that.

We've learned those tricks nowadays, and now we don't have that kind of a disease. But this long history of learning how not to fool ourselves—of having utter scientific integrity—is, I'm sorry to say, something that we haven't specifically included in any particular course that I know of. We just hope you've caught on by osmosis.

The first principle is that you must not fool yourself—and you are the easiest person to fool. So you have to be very careful about that. After you've not fooled yourself, it's easy not to fool other scientists. You just have to be honest in a conventional way after that.

I would like to add something that's not essential to the science, but something I kind of believe, which is that you should not fool the layman when you're talking as a scientist. I am not trying to tell you what to do about cheating on your wife, or fooling your girlfriend, or something like that, when you're not trying to be a scientist, but just trying to be an ordinary human being. We'll leave those problems up to you and your rabbi. I'm talking about a specific, extra type of integrity that is not lying, but bending over backwards to show how you are maybe wrong, that you ought to have when acting as a scientist. And this is our responsibility as scientists, certainly to other scientists, and I think to laymen.

For example, I was a little surprised when I was talking to a friend who was going to go on the radio. He does work on cosmology and astronomy, and he wondered how he would explain what the applications of this work were. "Well," I said, "there aren't any."

He said, "Yes, but then we won't get support for more research of this kind." *I* think that's kind of dishonest. If you're representing yourself as a scientist, then you should explain to the layman what you're doing—and if they don't want to support you under those circumstances, then that's their decision.

One example of the principle is this: If you've made up your mind to test a theory, or you want to explain some idea, you should always decide to publish it whichever way it comes out. If we only publish results of a certain kind, we can make the argument look good. We must publish *both* kinds of results.

I say that's also important in giving certain types of government advice. Supposing a senator asked you for advice about whether drilling a hole should be done in his state; and you decide it would be better in some other state. If you don't publish such a result, it seems to me you're not giving scientific advice. You're being used. If your answer happens to come out in the direction the government or the politicians like, they can use it as an argument in their favor; if it comes out the other way, they don't publish it at all. That's not giving scientific advice.

Other kinds of errors are more characteristic of poor science. When I was at Cornell, I often talked to the people in the psychology department. One of the students told me she wanted to do an experiment that went something like this—it had been found by others that under certain circumstances, *X*, rats did something, *A*. She was curious as to whether, if she changed the circumstances to *Y*, they would still do *A*. So her proposal was to do the experiment under circumstances *Y* and see if they still did *A*.

I explained to her that it was necessary first to repeat in her laboratory the experiment of the other person—to do it under condition *X* to see if she could also get result *A*, and then change to *Y* and see if *A* changed. Then she would know that the real difference was the thing she thought she had under control.

She was very delighted with this new idea, and went to her professor. And his reply was, no, you cannot do that, because the experiment has already been done and you would be wasting time. This was in about 1947 or so, and it seems to have been the general policy then to not try to repeat psychological experiments, but only to change the conditions and see what happens.

Nowadays there's a certain danger of the same thing happening, even in the famous field of physics. I was shocked to hear of an experiment done at the big accelerator at the National Accelerator Laboratory, where a person used deuterium. In order to compare his heavy hydrogen results to what might happen with light hydrogen he had to use data from someone else's experiment on light hydrogen, which was done on different apparatus. When asked why, he said it was because he couldn't get time on the program (because there's so little time and it's such expensive apparatus) to do the experiment with light hydrogen on this apparatus because there wouldn't be any new result. And so the men in charge of programs at NAL are so anxious for new results, in order to get more money to keep the thing going for public relations purposes, they are destroying—possibly—the value of the experiments themselves, which is the whole purpose of the thing. It is often hard for the experimenters there to complete their work as their scientific integrity demands.

All experiments in psychology are not of this type, however. For example, there have been many experiments running rats through all kinds of mazes, and so on—with little clear result. But in 1937 a man named Young did a very interesting one. He had a long corridor with doors all along one side where the rats came in, and doors along the other side where the food was. He wanted to see if he could train the rats to go in at the third door down from wherever he started them off. No. The rats went immediately to the door where the food had been the time before.

The question was, how did the rats know, because the corridor was so beautifully built and so uniform, that this was the same door as before? Obviously there was something about the door that was different from the other doors. So he painted the doors very carefully, arranging the textures on the faces of the doors exactly the same. Still the rats could tell. Then he thought maybe the rats were smelling the food, so he used chemicals to change the smell after each run. Still the rats could tell. Then he realized the rats might be able to tell by seeing the lights and the arrangement in the laboratory like any commonsense person. So he covered the corridor, and still the rats could tell.

He finally found that they could tell by the way the floor sounded when they ran over it. And he could only fix that by putting his corridor in sand. So he covered one after another of all possible clues and finally was able to fool the rats so that they had to learn to go in the third door. If he relaxed any of his conditions, the rats could tell.

Now, from a scientific standpoint, that is an A-number-one experiment. That is the experiment that makes rat-running experiments sensible, because it uncovers the clues that the rat is really using—not what you think it's using. And that is the experiment that tells exactly what conditions you have to use in order to be careful and control everything in an experiment with rat-running.

I looked into the subsequent history of this research. The next experiment, and the one after that, never referred to Mr. Young. They never used any of his criteria of putting the corridor on sand, or being very careful. They just went right on running rats in the same old way, and paid no attention to the great discoveries of Mr. Young, and his papers are not referred to, because he didn't discover anything about the rats. In fact, he discovered *all* the things you have to do to discover something about rats. But not paying attention to experiments like that is a characteristic of cargo cult science.

Another example is the ESP experiments of Mr. Rhine, and other people. As various people have made criticisms—and they themselves have made criticisms of their own experiments—they improve the techniques so that the effects are smaller, and smaller, and smaller until they gradually disappear. All the parapsychologists are looking for some experiment that can be repeated—that you can do again and get the same effect—statistically, even. They run a million rats no, it's people this time they do a lot of things and get a certain statistical effect. Next time they try it they don't get it any more. And now you find a man saying that it is an irrelevant demand to expect a repeatable experiment. This is *science*?

This man also speaks about a new institution, in a talk in which he was resigning as Director of the Institute of Parapsychology. And, in telling people what to do next, he says that one of the things they have to do is be sure they only train students who have shown their ability to get PSI results to an acceptable extent—not to waste their time on those ambitious and interested students who get only chance results. It is very dangerous to have such a policy in teaching—to teach students only how to get certain results, rather than how to do an experiment with scientific integrity.

So I have just one wish for you—the good luck to be somewhere where you are free to maintain the kind of integrity I have described, and where you do not feel forced by a need to maintain your position in the organization, or financial support, or so on, to lose your integrity. May you have that freedom. May I also give you one last bit of advice: Never say that you'll give a talk unless you know clearly what you're going to talk about and more or less what you're going to say.

27

Pathological Science
Irving Langmuir

Irving Langmuir (January 31, 1881 – August 16, 1957)—was an American chemist whose research spanned a number of fields. Working at the General Electric research laboratories in Schenectady, New York he discovered adding nitrogen to light bulbs could extend filament life, coined the term 'plasma' in reference to ionized gases, won a Nobel Prize in Chemistry in 1932 for his work on surface chemistry, and attempted weather control technology (Project Matterhorn). This lecture on Pathological Science is well-known for its advice on avoiding pseudoscience amongst scientific investigators.

Preface (by R.N. Hall)

On December 18, 1953, Dr. Irving Langmuir gave a colloquium at the Research Laboratory that will long be remembered by those in his audience. The talk was concerned with what Langmuir called "the science of things that aren't so," and in it he gave a colorful account of several examples of a particular kind of pitfall into which scientists may sometimes stumble.

Langmuir never published his investigations into the subject of Pathological Science. A tape recording was made of his speech, but this has been lost or erased. Recently, however, a microgroove disk transcription that was made from this tape was found among the Langmuir papers in the Library of Congress, This disk recording is of poor quality, but most of what he said can be understood with a little practice, and it constitutes the text of this report.

A small amount of editing was felt to be desirable. Some abortive or repetitious sentences were eliminated. Figures from corresponding publications were used to represent his blackboard

---

*Based on the colloquium at The Knolls Research Laboratory, December 18, 1953. Transcribed and edited by R. N. Hall and printed in GE Report No. 68-C-035 in April 1968. The report was made available online by Theo Pavlidis and Ken Steiglitz

sketches, and some references were added for the benefit of anyone wishing to undertake a further investigation of this subject. The disk recording has been transcribed back onto tape, and a copy is on file in the Whitney Library.

Gratitude is hereby expressed to the staff of the Manuscript Division of the Library of Congress for their cooperation in lending us the disk recording so we could obtain the best possible copy of the Langmuir speech, and for providing access to other related Langmuir papers.

Colloquium on Pathological Science

This is recorded by Irving Langmuir on March 8, 1954. It is transcribed from a tape recording, section number three, of the lecture on "Pathological Science" that I gave on December 18, 1953.

Davis-Barnes Effect

The thing started in this way. On April the 23rd 1929, Professor Bergen Davis from Columbia University came up and gave a colloquium in this Laboratory, in the old building, and it was very interesting. He told Dr. Whitney, and myself, and a few others something about what he was going to talk about beforehand and he was very enthusiastic about it and he got us interested in it, and well, I'll show you right on this diagram what kind of thing happened (Fig. 1).

He produced a beam of alpha rays from polonium in a vacuum tube. He had a parabolic hot cathode electron emitter with a hole in the middle, and the alpha rays came through it and could be counted by scintillations on a zinc sulfide screen with a microscope over here (Y and Z). The electrons were focused on this plate, so that for a distance there was a stream of electrons moving along with the alpha particles. Now you could accelerate the electrons and get them up to the velocity of the alpha particles. To get an electron to move with that velocity takes about 590 volts; so if you put 590 volts here, accelerating the electrons, the electrons would travel along with the alpha particles and the idea of the experiment was that if they moved along together at the same velocity they might recombine so that the alpha particle would lose one of its charges, would pick up an electron, so that instead of being a helium atom with two positive charges it would only have one charge. Well, if an alpha particle with a double charge had one electron, it's like the Bohr theory of the hydrogen atom, and you know its energy levels. It's just like a hydrogen atom, with a Balmer series, and you can calculate the energy necessary to knock off this electron and so on.

Well, what they found, Davis and Barnes, was that if this velocity was made to be the same as that of the alpha particle there was a loss in the number of deflected particles. If there were no electrons, for example, and no magnetic field, all the alpha particles would be collected over here (Y) and they had something of the order of 50 per minute which they counted over here. Now if you put on a magnetic field you could deflect the alpha particles so they go down here (Z). But if they picked up an electron then they would only have half the charge and therefore they would only be deflected half as much and they would not strike the screen.

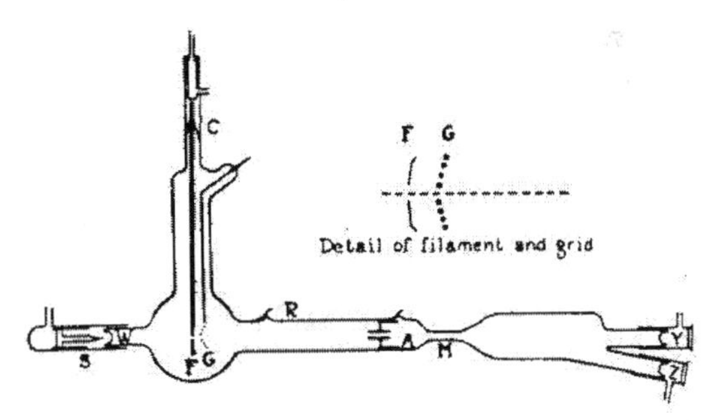

Figure 1 Diagram of first experimental tube. S, radioactive source; W, thin glass window; F, filament; G, grid; R, lead to silvered surface; A, second anode; M, magnetic field; C, copper seals; Y, and Z, zinc sulfide screens.

Now the results that they got, or said they got at that time, were very extraordinary. They found that not only did these electrons combine with the alpha particles when the electron velocity was 590 volts, but also at a series of discrete differences of voltage. When the velocity of the electrons was less or more than that velocity by perfectly discrete amounts, then they could also combine. All the results seemed to show that about 80' of them combined.

In other words, there was about an 80' change in the current when the conditions were right. Then they found that the velocity differences had to be exactly the velocities that you can calculate from the Bohr theory. In other words, if the electron coming along here happened to be going with a velocity equal to the velocity that it would have if it was in a Bohr orbit, then it will be captured.

Of course, that makes a difficulty right away because in the Bohr theory when there is an electron coming in from infinity it has to give up half its energy to settle into the Bohr orbit. Since it must conserve energy, it has to radiate out, and it radiates out an amount equal to the energy that it has left in the orbit So, if the electron comes in with an amount of energy equal to the amount you are going to end up with, then you have to radiate an amount of energy equal to twice that, which nobody had any evidence for. So there was a little difficulty which never was quite resolved although there were two or three people including some in Germany who worked up theories to account for how that might be. Sommerfeld, for example, in Germany. He worked up a theory to account for how the electron could be captured if it had a velocity equal to what it was going to have after it settled down into the orbit.

Well, there were these discrete peaks, each one corresponding to one of the energy levels in the Bohr theory of the helium atom, and nothing else. Those were the only things they recorded. So you had these discrete peaks. Well, how wide were they? Well, they were one *hundredth of a volt wide.* In other words, you had to have 590 volts. That would give you equal velocities but there were other peaks, and I think the next velocity would be about 325.1 volts. If you had that voltage, then you got beautiful capture. If you didn't, if you changed it by one hundredth of a volt—nothing. It would go right from 80% down to nothing. It was sharp. They were only able to measure to a hundredth of a volt so it was an all-or-none effect. Well, besides this peak at this point, there were ten or twelve different lines in the Balmer series, all of which could be detected, and all of which had an 80% efficiency. (See Fig. 2.) They almost completely captured all the electrons when you got exactly on the peak.

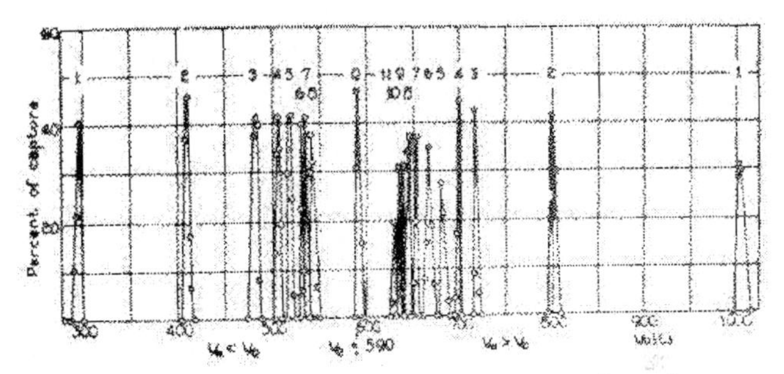

Figure 2 Electron capture as a function of accelerating voltage. [Copy from Barnes. Phys. Rev., 35, 217 (1930).]

Well, in the discussion, we questioned how, experimentally, you could examine the whole spectrum; because each count, you see, takes a long time. There was a long series of alpha particle counts, that took two minutes at a time, and you had to do it ten or fifteen times and you had to adjust the voltage to a hundredth of a volt. If you have to go through steps of a hundredth of a volt each and to cover all the range from 330 up to 900 volts, you'd have quite a job. (Laughter) Well, they said that they didn't do it quite that way. They had found by some preliminary work that they did check with the Bohr orbit velocities so they knew where to look for them. They found them sometime not exactly where they expected them but they explored around in that neighborhood and the result was that they got them with extraordinary precision. So high, in fact, that they were sure they'd be able to check the Rydberg constant more accurately than it can be done by studying the hydrogen spectrum, which is something like one in 10^8 At any rate, they had no inhibitions at all as to the accuracy which could be obtained by this method especially since they were measuring these voltages within a hundredth of a volt.

Anybody who looks at the setup would be a little doubtful about whether the electrons had velocities that were fixed and definite within 1/100 of a volt because this is not exactly a homogeneous field. The distance was only about 5 mm in which they were moving along together.

Well, in his talk, a few other things came out that were very interesting. One was that the percent-age of capture was always around 80%. The curves would come along like this as a function of voltage (Fig. 2). The curve would come along at about 80% and there would be a sharp peak up here and another sharp peak here and, well, all the peaks were about the same height.

Well, we asked, how did this depend upon current density? "That's very interesting," he said, "It doesn't depend at all upon current density."

We asked, "How much could you change the temperature of the cathode here?"

"Well," he said, "that's the queer thing about it. You can change it all the way down to room temperature." (Laughter)

"Well," I said, "then you wouldn't have any electrons.'

"Oh, yes," he said, "if you check the Richardson equation and calculate, you'll find that you get electrons even at room temperature and those are the ones that are captured."

'Well," I said, "there wouldn't be enough to combine with all the alpha particles and, besides that, the alpha particles are only there for a short time as they pass through and the electrons are a long way apart at such low current densities, at 10^{-20} amperes or so." (Laughter)

He said, "That seemed like quite a great difficulty. "But," he said, "you see it isn't so bad because we now know that the electrons are waves. So the electron doesn't have to be there at all in order to combine with something. Only the waves have to be there and they can be of low intensity and the quantum theory causes all the electrons to pile in at just the right place where they are needed." So he saw no difficulty. And so it went.

Well, Dr. Whitney likes the experimental method, and these were experiments, very careful experiments, described in great detail, and the results seemed to be very interesting from a theoretical point of view. So Dr. Whitney suggested that he would like to see these experiments repeated with a Geiger counter instead of counting scintillations, and C.W. Hewlett, who was here working on Geiger counters, had a setup and it was proposed that we would give him one of these, maybe at a cost of several thousand dollars or so for the whole equipment, so that he could get better data. But I was a little more cautious. I said to Dr. Whitney that before we actually give it to him and just turn it over to him. it would be well to go down and take a look at these experiments and see what they really mean. Well, Hewlett was very much interested and I was interested so only about two days later, after this colloquium, we went down to New York. We went to Davis's Laboratory at Columbia University, end we found that they were very glad to see us, very proud to show us all their results, so we started in early in the morning.

We sat in the dark room for half an hour to get our eyes adapted to the darkness so that we could count scintillations. I said, first I would like to see these scintillations with the field on and with the field off. So I looked in and I counted about 50 or 60. Hewlett counted 70, and I counted somewhat lower. On the other hand, we both agreed substantially. What we found was this. These scintillations were quite bright with your eyes adapted, and there was no trouble at all about counting them, when these alpha particles struck the screen. They came along at a rate of about 1 per second. When you put on a magnetic field and deflected them out, the count came down to about 17, which was a pretty high percentage, about 25% background. Barnes was sitting with us, and he said that's probably radioactive contamination of the screen. Then, Barnes counted and he got 230 on the first count and about 200 on the next, and when he put on the field it went down to about 25. Well, Hewlett and I didn't know what that meant but we couldn't see 230. Later, we understood the reason.

I had seen, and we discussed a little at that point, that the eyepiece was such that as you looked through, you got some flashes of light which I took to be flashes that were just outside the field of view that would give a diffuse glow that would be perceptible. And you could count them as events. They clearly were not particles that struck the screen where you saw it, but nevertheless, they seemed to give a dif-fuse glow and they came at discrete intervals and you could count those if you wanted. Well, Hewlett counted those too and I didn't. That accounted for some difference. Well, we didn't bother to check into this, and we went on.

Well, I don't want to spend too much time on this experiment. I have a 22-page letter that I wrote about these things and I have a lot of notes. The gist of it was this. There was a long table at which Barnes was sitting, and he had another table over here where he had an assistant of his named Hull who sat here looking at a big scale voltmeter, or potentiometer really, but it had a scale that went from one to a thousand volts and on that scale that went from one to a thousand, he read hundredths of a volt. (Laughter) He thought he might be able to do a little better than that. At any rate, you could interpolate and put down figures, you know. Now the room was dark except for a little light here on which you could read the scale on that meter. And it was dark except for the dial of a clock and he counted scintillations for two minutes.

He said he always counted for two minutes. Actually, I had a stop watch and I checked him up. They sometimes were as low as one minute and ten seconds and sometimes one minute and fifty-five seconds but he counted them all as two minutes, and yet the results were of high accuracy!

Well, we made various suggestions. One was to turn off the voltage entirely. Well, then Barnes got some low values around 20 or 30, or sometimes as high as 50. Then to get the conditions on a peak he adjusted the voltage to two hundred and, well some of those readings are interesting; 325.01. That's the figure I put down, and there he got only a reading of 52, whereas before when he was on the peak, he got about. He didn't like that very much 80 he tried changing this to .02; a change of one hundredth of a volt. And there he got 48. *Then he went in between.* (Laughter) They fell off, you see, so he tried 325.015 and then he got 107. So that was a peak.

Well, a little later, I whispered to Hull who was over here adjusting the voltage, holding it constant, I suggested to him to make it one tenth of a volt different. Barnes didn't know this and he got 96. Well, when I suggested this change to Hull, you could see immediately that he was amazed. He said, "Why, that's too big a change. That will put it way off the peak." That was almost one tenth of a volt, you see, Later I suggested taking a whole volt. (Laughter)

Then we had lunch. We sat for half an hour in the dark room so as not to spoil our eyes and then we had some readings at zero volts and then we went back to 325.03. We changed by one hundredth of a volt and there he got 110. And now he got two or three readings at 110.

And then I played a dirty trick. I wrote out on a card of paper 10 different sequences of V and 0. I meant to put on a certain voltage and then take it off again. Later I realized that that wasn't quite right because when Hull took off the voltage, he sat back in his chair—there was nothing to regulate at zero, so he didn't. Well, of course, Barnes saw him whenever he sat back in his chair. Although, the light wasn't very bright, he could see whether he was sitting back in his chair or not so he knew the voltage wasn't on and the result was that he got a corresponding result. So later I whispered, "Don't let him know that you're not reading," and I asked him to change the voltage from 325 down to 320 so he'd have something to regulate and I said, "Regulate it just as carefully as if you were sitting on a peak." So he played the part from that time on, and from that time on Barnes' readings had nothing whatever to do with the voltages that were applied. Whether the voltage was at one value or another didn't make the slightest difference. After that he took twelve readings, of which about half of them were right and the other half were wrong, which was about what you would expect out of two sets of values.

I said, "You're through. You're not measuring anything at all. You never have measured anything at all."

"Well," he said, "the tube was gassy. (Laughter) The temperature has changed and therefore the nickel plates must have deformed themselves so that the electrodes are no longer lined up properly."

"Well," I said, "isn't this the tube in which Davis said he got the same results when the filament was turned off completely?"

"Oh, yes," he said, "but we always made blanks to check ourselves, with and without the voltage on."

He immediately—without giving any thought to it—he immediately had an excuse. He had a reason for not paying any attention to any wrong results. It just was built into him. He just had worked that way all along and always would. There is no question but what he is honest; he *believed* these things, absolutely.

Hewlett stayed there and continued to work with him for quite a while and I went in and talked it over with Davis and he was simply dumbfounded. He couldn't believe a word of it. He said, "It absolutely can't be," he said. "Look at the way we *found* those peaks before we knew anything about the Bohr theory. We took those values and calculated them up and they checked exactly. Later on, after we got confirmation, in order to save time, to see whether the peaks were there we would calculate ahead of time." He was so sure from the whole history of the thing that it was utterly impossible that there never had been any measurements at all that he just wouldn't believe it.

Well, he had just read a paper before the Research Laboratory at Schenectady, and he was going to read the paper the following Saturday before the National Academy of Sciences; which he did, and gave the whole paper. And he wrote me that he was going to do so on the 24th. I wrote to him on the day after I got back. Our letters crossed in the mails and he said that he had been thinking over the various things that I had told him, and his confidence wasn't shaken, so he went ahead and presented the paper before the National Academy of Sciences.

Then I wrote him a 22-page letter giving all our data and showing really that the whole approach to the thing was wrong; that he was counting hallucinations, which I find is common among people who work with scintillations if they count for too long. Barnes counted for six hours a day and it never fatigued him. Of course it didn't fatigue him, because it was all made up out of his head. (Laughter) He told us that you mustn't count the bright particles. He had a beautiful reason for why you mustn't pay any attention to the bright flashes. When Hewlett tried to check his data he said, "Why, you must be counting those bright flashes. Those things are only due to radioactive contamination or something else." He had a reason for rejecting the very essence of the thing that was important. So I wrote all this down in this letter and I got no response, no encouragement. For a long time Davis wouldn't have anything to do with it. He went to Europe for a six months leave of absence, came back later, and I took up the matter with him again [1].

In the meantime, I sent a copy of the letter that I had written to Davis to Bohr asking him to hold it confidential but to pass it on to various people who would be trying to repeat these experiments. To Professor Sommerfeld and other people and it headed off a lot of experimental work that would have gone on. And from that time on, nobody ever made another experiment except one man in England who didn't know about the letter that I had written to Bohr [2]. And he was not able to confirm any of it. Well, a year and a half later, in 1931, there was just a short little article in the Physical Review in which they say that they haven't been able to reproduce the effect [3]. "The results reported in the earlier paper depended upon observations made by counting scintillations visually. The scintillations produced by alpha particles on a zinc sulfide screen are a threshold phenomenon. It is possible that the number of counts may be influenced by external suggestion or auto-suggestion to the observer," and later in that paper they said that they had not been able to check any of the older data. And they didn't even say that the tube was gassy. (Laughter)

To me, the thing is extremely interesting, that men, perfectly honest, enthusiastic over their work, can so completely fool themselves. Now what was it about that work that made it so easy for them to do that? Well, I began thinking of other things. I had seen K W. Wood and told him about this phenomenon because he's a good experimenter and doesn't make such mistakes himself very often, if at all. And he told me about the N-rays that he had an experience with back in 1904. So I looked up the data on the N-rays [4,5].

N-rays

In 1903, Blondlot, who was a well-thought-of French scientist, member of the Academy of Sciences, was experimenting with x-rays as almost everybody was in those days, The effect that he observed was something of this sort. I won't give the whole of it, I'll just give a few outstanding points. He found that if you have a hot wire, a platinum wire, or a Nernst filament or anything that's heated very hot inside an iron tube and you have a window cut in it and you have a piece of aluminum about 1/8 of an inch thick on it, that some rays come out through that aluminum window. Oh, it can be as much as two or three inches thick and go through aluminum, these rays can, but not through iron. The rays that come out of this little window fall on a faintly illuminated object, so that you can just barely see it. You must sit in a dark room for a long time and he used a calcium sulfide screen which can be illuminated with light and gave out a very faint glow which could be seen in a dark room. Or he used a source of light from a lamp shining through a pinhole and maybe through another pinhole so as to get a faint light on a white surface that was just barely visible.

Now he found that if you turn this lamp on so that these rays that come out of this little aluminum slit would fall on this piece of paper that you are looking at, you could see it *much better*. Oh, much better, and therefore you could tell whether the rays would go through or not. He said later that a great deal of skill is needed. He said you mustn't ever look at the source. You don't look directly at it. He said that would tire your eyes. Look away from it, and he said pretty soon you'll see it, or you don't see it, depending on whether the N-rays are shining on this piece of paper. In that way, you can detect whether or not the N-rays are acting.

Well, he found that N-rays could be stored up in things. For example, you could take a brick. He found that N-rays would go through black paper and would go through aluminum. So he took some black paper and wrapped a brick up in it and put it out in the street and let the sun shine through the black paper into the brick and then he found that the brick would store N-rays and give off the N-rays even with the black paper on it. He would bring it into the laboratory and you then hold that near the piece of paper that you're looking at, faintly illuminated, and you can see it much more accurately. Much better, if the N-rays are there, but not if it's too far away. Then, he would have very faint strips of phosphorescent paint and would let a beam of N-rays from two slits come over and he would find exactly where this thing intensified its beam.

Well, you'd think he'd make such experiments as this. To see if with ten bricks you got a stronger effect than you did with one. No, not at all. He didn't get any stronger effect. It didn't do any good to increase the intensity of the light. You had to depend upon whether you could see it or whether you couldn't see it. And there, the N-rays were very important.

Now, a little later, he found that many kinds of things gave off N-rays. A human being gave off N-rays, for example. If someone else came into the room, then you probably could see it. He also found that if someone made a loud noise that would spoil the effect. You had to be silent. Heat, however, increased the effect, radiant heat. Yet that wasn't N-rays itself. N-rays were not heat because heat wouldn't go through aluminum. Now he found a very interesting thing about it was that if you take the brick that's giving off N-rays and hold it close to your head it goes through your skull and it allows you to see the paper better. Or you can hold the brick near the paper, that's all right too.

Now he found that there were some other things that were like negative N-rays. He called them N'-rays. The effect of the N'-rays is to decrease the visibility of a faintly illuminated slit. That works too, but only if the angle of incidence is right. If you look at it tangentially you find that the thing in-creases the intensity when you look at it from this point of view. It decreases if you look at it normally and it increases if you look at it tangentially. All of which is very interesting. And he published many papers on it. One right after the other and other people did too, confirming Blondlot's results. And there were lots of papers published and at one time about half of them that were confirming' the results of Blondlot. You see, N-rays ought to be important because x-rays were known to be important and alpha rays were, and N-rays were somewhere in between so N-rays must be very important. (Laughter)

Well, R.W. Wood heard about these experiments—everybody did more or less. So R.W. Wood went over there and at that time Blondlot had a prism, quite a large prism of aluminum, with a 60° angle and he had a Nernst filament with a little slit about 2 mm wide. There were two slits, 2 mm wide each. This beam fell on the prism and was refracted and he measured the refractive index to three significant figures. He found that it wasn't monochromatic, that there were several different components to the N-rays and he found different refractive indices for each of these components. He could measure three or four different refractive indices each to two or three significant figures, and he was repeating some of these and showing how accurately they were repeatable, showing it to R.W. Wood in this dark room.

Well, after this had gone on for quite a while, and Wood found that he was checking these results very accurately, measuring the position of the little piece of paper within a tenth of a millimeter although the slits were 2 mm wide, and Wood asked him about that. He said, "How? How could you, from just the optics of the thing, with slits two millimeters wide, how do you get a beam so fine that you can detect its position within a tenth of a millimeter?"

Blondlot said, "That's one of the fascinating things about the N-rays. They don't follow the ordinary laws of science that you ordinarily think of." He said, "You have to consider these things all by themselves. They are very interesting, but you have to discover the laws that govern them."

Well, in the meantime, the room being very dark, Wood asked him to repeat some of these measurements which he was only too glad to do. But in the meantime, R.W. Wood put the prism in his pocket and the results checked perfectly with what he had before. (Laughter) Well, Wood rather cruelly published that [6,7]. And that was the end of Blondlot.

Nobody accounts for by what methods he could reproduce those results to a tenth of a millimeter. Wood said that he seemed to be able to do it but no-body understands that. Nobody understands lots of things. But some of the Germans came out later—Pringsheim was one of them—came out with an extremely interesting story. They had tried to repeat some of Blondlot's experiments and had found this. One of the experiments was to have a very faint source of light on a screen of paper and to make sure that you are seeing the screen of paper you hold your hand up like this and move it back and forth. And if you can see your hand move back and forth then you know it is illuminated. One of the experiments that Blondlot made was that the experiment was made much better if you had some N-rays falling on the piece of paper. Pringsheim was repeating these in Germany and he found that if you didn't know where the paper was, whether it was here or here (in front or behind your hand), it worked just as well. That is, you could see your hand just as well if you held it back of the paper as if you held it in front of it. Which is the natural thing, because this is a threshold phenomenon. And a threshold phenomenon means that you don't know, *you really don't know*, whether you are seeing it or not. But if you have your hand there, well, of course, you see your hand because you *know* your hand's there, and that's just enough to win you over to where you know that you see it. But you know it just as well if the paper happens to be in front of your hand instead of in back of your hand, because you don't know where the paper is but you *do* know where your hand is. (Laughter)

Mitogenetic Rays

Well, let's go on. About 1923, there was a whole series of papers by Gurwitsch and others. There were hundreds of them published on mitogenetic rays [8]. There are still a few of them being published. I don't know how many of you have ever heard of mitogenetic rays. They are rays that are given off by growing plants, living things, and they were proved, according to Gurwitsch, that they were something that would go through quartz but not through glass. They seemed to be some sort of ultraviolet light.

The way they studied these was this. You had some onion roots—onions growing in the dark or in the light and the roots will grow straight down. Now if you had another onion root nearby, and this onion root was growing down through a tube or something, going straight down, and another onion root came nearby, this would develop so that there were more cells on one side than the other. One of the tests they had made at first was that this root would bend away. And as it grew this would change in direction which was evidence that something had traveled from one onion root to the other. And if you had a piece of quartz in between it would do it, but if you put glass in between it wouldn't. So this radiation would not go through glass but it would go through quartz.

Well, it started in that way. Then everything gave off mitogenetic rays, anything that remotely had anything to do with living things. And then they started to use photoelectric cells to check it and whatever they did they practically always found that if you got the conditions just right, you could just detect it and prove it. But if you looked over those photographic plates that showed this ultraviolet light you found that the amount of light was not much bigger than the natural particles of the photographic plate so that people could have different opinions as to whether it did or didn't show this effect and the result was that less than half of the people who tried to repeat these experiments got any confirmation of it; and so it went. Well, I'll go on before I get too far along.

Characteristic Symptoms of Pathological Science

The characteristics of this Davis-Barnes experiment and the N-rays and the mitogenetic rays, they have things in common. These are cases where there is no dishonesty involved but where people are tricked into false results by a lack of understanding about what human beings can do to themselves in the way of being led astray by subjective effects, wishful thinking or threshold interactions. These are examples of pathological science. These are things that attracted a great deal of attention. Usually hundreds of papers have been published upon them. Sometimes they have lasted for fifteen or twenty years and then they gradually die away.

Now, the characteristic rules are these (see Table I)

TABLE I

Symptoms of Pathological Science:

1. The maximum effect that is observed is produced by a causative agent of barely detectable intensity, and the magnitude of the effect is substantially independent of the intensity of the cause.
2. The effect is of a magnitude that remains close to the limit of detectability; or, many measurements are necessary because of the very low statistical significance of the results.
3. Claims of great accuracy.
4. Fantastic theories contrary to experience.
5. Criticisms are met by *ad hoc* excuses thought up on the spur of the moment.
6. Ratio of supporters to critics rises up to somewhere near 50% and then falls gradually to oblivion.

The maximum effect that is observed is produced by a causative agent of barely detectable intensity. For example, you might think that if one onion root would affect another due to ultraviolet light, you'd think that by putting on an ultraviolet source of light you could get it to work better. Oh no! OH NO! It had to be just the amount of intensity that's given off by an onion root. Ten onion roots wouldn't do any better than one and it doesn't make any difference about the distance of the source. It doesn't follow any inverse square law or anything as simple as that, and so on. In other words, *the effect is independent of the intensity of the cause*. That was true in the mitogenetic rays, and it was true in the N-rays. Ten bricks didn't have any more effect than one. It had to be of low intensity. We know why it had to be of low intensity: so that you could fool yourself so easily. Otherwise, it wouldn't work. Davis-Barnes worked just as well when the filament was turned off. They counted scintillations.

Another characteristic thing about them all is that, these observations are near *the threshold of visibility of the eyes*. Any other sense, I suppose, would work as well. *Or many measurements are necessary, many measurements because of very low statistical significance of the results*. In the mitogenetic rays particularly it started out by seeing something that was bent. Later on, they would take a hundred onion roots and expose them to something and they would get the average position of all of them to see whether the average had been affected a little bit by an appreciable amount. Or statistical measurements of a very small effect which by taking large numbers were thought to be significant. Now the trouble with that is this. There is a habit with most people, that when measurements of low significance are taken they find means of rejecting data. They are right at the threshold value and there are many reasons why you can discard data. Davis and Barnes were doing that right along. If things were doubtful at all why they would discard them or not discard them depending on whether or not they fit the theory. They didn't know that, but that's the way it worked out.

There are claims of great accuracy. Barnes was going to get the Rydberg constant more accurately than the spectroscopists could. Great sensitivity or great specificity, we'll come across that particularly in the Allison effect.

Fantastic theories contrary to experience. In the Bohr theory, the whole idea of an electron being captured by an alpha particle when the alpha particles aren't there just because the waves are there doesn't make a very sensible theory.

Criticisms are met by ad hoc excuses thought up on the spur of the moment. They always had an answer—always.

The ratio of the supporters to the critics rises up somewhere near 50% and then falls gradually to oblivion. The critics can't reproduce the effects. Only the supporters could do that. In the end, nothing was salvaged. Why should there be? There isn't anything there. There never was. That's characteristic of the effect. Well, I'll go quickly on to some of the other things.

Allison Effect

The Allison effect is one of the most extraordinary of all [9]. It started in 1927. There were hundreds of papers published in the American Physical Society, the Physical Review, the Journal of the American Chemical Society—hundreds of papers. Why, they discovered five or six different elements that were listed in the Discoveries of the Year. There were new elements discovered—Alabamine, Virginium, a whole series of elements and isotopes were discovered by Allison.

The effect was very simple. There is the Faraday effect by which a beam of polarized light passing through a liquid which is in a magnetic field is rotated—the plane of polarization is rotated by a longitudinal magnetic field. Now that idea has been known for a long time and it has a great deal of importance in connection with light shutters. At any rate, you can let light through or not depending upon the magnetic field. Now the experiment of Allison's was this (Fig. 3). They had a glass cell and a coil of wire around it (B1, B2) and you have wires coming up here, a Lecher system. Here you have a spark gap, so a flash of light comes through here and goes through a Nicol prism over here and another one over here, and you adjust this one with a liquid like water or carbon disulfide or something like that in the cell so that there was a steady light over here. If you have a beam of light and you polarize it and then you turn on a magnetic field, why you see that you could rotate the plane of polarization. There will be an increase in the brightness of the light when you put a magnetic field on here. Now they wanted to find the time delay, how long it takes. So they had a spark and the same field that produced the spark induced a current through the coil, and by sliding this wire along the trolley of the Lecher system, they could cause a compensating delay. The sensitivity of this thing was so great that they could detect differences of about 3 x 10^{-10} seconds. By looking in here they could see these flashes of light, the light from the sparks, and they tried to decide as they changed the position of this trolley whether it got brighter or dimmer and they set it for a minimum, and measured the position of the trolley. They put in here—in this glass tube—they put a water solution and added some salt to it. And they found that the time lag was changed, so that they got a change in the time lag depending upon the presence of salts.

Now they first found—very quickly—that if you put in a thing like ethyl alcohol that you got one characteristic time lag, and with acetic acid another one, quite different. But if you had ethyl acetate you got the sum of the two. You got two peaks. So that you could analyze ethyl acetate and find the acetic acid and the

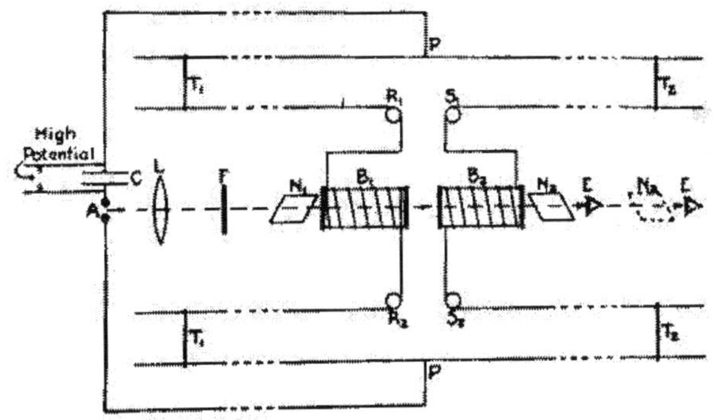

Figure 3 Diagram of apparatus and connections. [Copy from F. Allison, Phys. Rev., 30, 66 (1927). Fig. 1]

ethyl alcohol. Then they began to study salt solutions and they found that only the metal elements counted but they didn't act as an ion. That is, all potassium ions weren't the same, but potassium nitrate and potassium chloride and potassium sulfate all had quite characteristic different points, that were a characteristic of the compound. It was only the positive ion that counted and yet the negative ions had a modifying effect. But you couldn't detect the negative ions directly.

Now they began to see how sensitive it was. Well, they found that any intensity more than about 10^{-8} molar solution would always produce the maximum effect, and you'd think that that would be kind of discouraging from the analytical point of view, but no, not at all. And you could make quantitative measurements to about three significant figures by diluting the solutions down to a point where the effect disappeared. Apparently, it disappeared quite sharply when you got down to about 10^{-8} or 3. 42 x 10^{-8} in concentration, or something of that sort and then the effect would disappear. Otherwise, you would get it, so that you could detect the limit within this extraordinary degree of accuracy.

Well, they found that things were entirely different, even in these very dilute solutions, in sodium nitrate from what it was with sodium chloride. Nevertheless, it was a characteristic which depended upon the compound even though the compound was dis-associated into ions at those concentrations. That didn't make any difference but it was fact that was experimentally proven. They then went on to find that the isotopes all stick right out like sore thumbs with great regularity. In the case of lead, they found sixteen isotopes. These isotopes were quite regularly spaced so that you could get 16 different positions and you could assign numbers to those so that you can identify them and tell which they are. Unfortunately, you couldn't get the concentrations quantitatively, even the dilution method didn't work quite right because they weren't all equally sensitive. You could get them relatively but only approximately. Well, it became important as a means of detecting elements that hadn't yet been discovered, like Alabamine and elements that are now known, and filling out the periodic table. All the elements in the periodic table were filled out that way and published.

But a little later, in 1945 or 46, I was at the University of California. Owen Latimer who is now Head of the Chemistry Department there—not Owen Latimer, Wendell Latimer—had had a bet with G.N. Lewis (in 1932). He said, "There's something funny about this Allison effect, how they can detect isotopes." He had known somebody who had been down with Allison and who had been very much impressed by the effect and he said to Lewis, "I think I'll go down and see Allison, to Alabama, and see what there is in it. I'd like to use some of these methods."

Now people had begun to talk about spectroscopic evidence that there might be traces of hydrogen of atomic weight three. It wasn't spoken of as tritium at that time but hydrogen of atomic weight three that might exist in small amounts. There was a little spectroscopic evidence for it and Latimer said, "Well, this might be a way of finding it. I'd like to be able to find it." So he went and spent three weeks at Alabama with Allison and before he went he talked it over with G.N. Lewis about what he thought the prospects were and Lewis said, "I'll bet you ten dollars you'll find that there's nothing in it." And so they had this bet on. He went down there and he came back. He set up the apparatus and made it work so well that G.N. Lewis paid him the ten dollars. (Laughter) He then discovered tritium and he published an article in the Physical Review [10]. Just a little short note saying that using Allison's method he had detected the isotope of hydrogen of atomic weight three. And he made some sort of estimate as to its concentration.

Well, nothing more was heard about it. I saw him then, seven or eight years after that. I had written these things up before, about this Allison effect, and I told him about this point of view and how the Allison effect fits all these characteristics. Well, I know at that time at one of the meetings of the American Chemical Society there was great discussion as to whether to accept papers on the Allison effect. There they decided: No, they would not accept any more papers on the Allison effect, and I guess the Physical Review did too. At any rate, the American Chemical Society decided that they would not accept any more manuscripts on the Allison effect. However, after they had adopted that as a firm policy, they did accept one more a year or two later because here was a case where all the people in the faculty here had chosen twenty or thirty different solutions that they had made up and they had labeled them all secretly and they had taken every precaution to make sure that nobody knew what was in these solutions, and they had given them to Allison and he had used his method on them and he had gotten them all right, although many of them were at concentrations of 10^{-6} and so on, molar. That was sufficiently definite—good experimental methods—and it was accepted for publication by the American Chemical Society but that was the last [11]. You'd think that would be the beginning, not the end.

Anyway, Latimer said, "You know, I don't know what was wrong with me at that time,' He said, "After I published that paper

I never could repeat the experiments again. I haven't the least idea why." "But," he said, "Those results were wonderful, I showed them to G.N. Lewis and we both agreed that it was all right. They were clean cut. I checked myself every way I knew how to. I don't know what else I could have done, but later on I just couldn't ever do it again."

I don't know what it is. That's the kind of thing that happens in all of these. All the people who had anything to do with these things find that when you get through with them—you *can't* account for Bergen Davis saying that they didn't calculate those things from the Bohr theory, that they were found by empirical methods without any idea of the theory. Barnes made the experiments, brought them in to Davis, and Davis calculated them up and discovered all of a sudden that they fit the Bohr theory. He said Barnes didn't have anything to do with that. Well, take it or leave it, how did he do it? It's up to you to decide. I can't account for it. All I know is that there was nothing salvaged at the end, and therefore none of it was ever right, and Barnes never did see a peak. You can't have a thing halfway right.

Extrasensory Perception

Well, there's Rhine. I spent a day with Rhine at Duke University at the meeting of the American Chemical Society, probably about 1934. Rhine had published a book and I'll just tell you a few things. First of all, I went in and told Rhine these things. I told him the whole story. I said these things (Table I) are the characteristics of those things that aren't so. They are all characteristics of your thing too. (Laughter) He said, "I wish you'd publish that. I'd love to have you publish it. That would stir up an awful lot of interest." He said, "I'd have more graduate students. We ought to have more graduate students. This thing is so important that we should have more people realize its importance. This should be one of the biggest departments in the university."

Well, I won't tell you the whole story with Rhine, because I talked with him all day. He uses cards which you guess at by turning over. You have extra-sensory perception. You have 25 cards and you deal them out face down, or one person looks at them, and the other person on the other side of the screen looks at them and you read his mind. The other thing is for nobody to know what the cards are, in which case they are turned over without anybody looking at them. You record them and then you look them up and see if they check and that's telepathy, or clairvoyance rather. Telepathy is when you can read another person's mind.

Now a later form of the thing would be for you to decide now and write down what the cards are going to be when they are shuffled tomorrow. That works too. (Laughter)

All of these things are nice examples where the magnitude of the effect is entirely independent of magnitude of the cause. That is, the experiments worked just as well where the shuffling is to be done tomorrow as when it was done some time ago. It doesn't make any difference in the results. There is no appreciable difference between clairvoyance and telepathy. Although, if you try to think of the mechanisms of the two, it should be quite different. In order to get the cards to telegraph you all the information that's in them as to how they are arranged, and so on, when they are stacked up on top of each other and to have it given in the right sequence, it is rather difficult to think of a mechanism. On the other hand, it is conceivable that there may be some sort of mechanism in the brain that might send out some sort of unknown messages that could be picked up by some other brain. That's a different order of magnitude. A different order of difficulty. But they were all the same from Rhine's point of view.

Well, now, the little things that I have are these. There are many more I could give you. Rhine said being in quite a philosophical mood, "It's funny how the mind tries to trick you." He said, "People don't like these experiments. I've had millions of these cases where the average is about 7 out of 25." You'd expect 5 out of 25 to come right by chance and on the grand average they come out, oh, out of millions, or hundreds of millions of cases, they average around 7. Well, to get 7 out of 25 would be a common enough occurrence but if you take a large number and you get 7, well you doubt the statistics or the statistical application or, above all, what I think of and I want to give you reasons for thinking, is the rejection of a small percentage of the data.

I'll go first, before I get into what Rhine said, and say this: David Langmuir, a nephew of mine, who was in the Atomic Energy Commission, when he was with the Radio Corporation of America a few years ago, he and a group of other young men thought they would like to check up Rhine's work so they got some cards and they spent many evenings together finding how these cards turned up and they got well above 5. They began to get quite excited about it and they kept on, and they kept on, and they were right on the point of writing Rhine about the thing. And they kept on a little longer and things began to fall off, and fall off a little more, and they fell off a little more. And after many, many, many days, they fell down to an average of five—grand average—so they *didn't* write to Rhine. Now if Rhine had received that information, that this reputable body of men had gone ahead and gotten a value of 8 or 9 or 10 after so many trials, why he would have put it in his book How much of that sort of thing, when you are fed information of that sort by people who are interested—how are you going to weigh the things that are published in the book?

Now an illustration of how it works is this. He told me that, "People don't like me," he said "I took a lot of cards and sealed them up in envelopes and I put a code number on the outside, and I didn't trust anybody to know that code. Nobody!"

(A section of the speech is missing at this point. It evidently described some tests that gave scores below 5.). "...the idea of having this thing sealed up in the cards as though I didn't trust them, and therefore to spite me they made it purposely low."

"Well," I said, "that's interesting—interesting a lot, because you said that you'd published a summary of all of the data that you had. And it comes out to be 7. It is now within your power to take a larger percentage including those cards that are sealed up in those envelopes which could bring the whole thing back down to five. Would you do that?

"Of course not," he said. "That would be dishonest."

"Why would it be dishonest?"

"The low scores are just as significant as the high ones, aren't they? They *proved* that there's something there just as much, and therefore it wouldn't be fair."

I said, "Are you going to count them, are you going to reverse the sign and count them, or count them as credits?"

"No, No," he said.

I said, "What have you done with them? Are they in your book?"

"No."

"Why, I thought you said that all your values were in your book. Why haven't you put those in?"

"Well," he said, "I haven't had time to work them up."

"Well, you know all the results, you told me the results."

"Well," he said, "I don't give the results out until I've had time to digest them."

I said, "How many of these things have you?" He showed me filing cabinets—a whole row of them. Maybe hundreds of thousands of cards. He has a filing cabinet that contained nothing but these things that were done in sealed up envelopes. And they were the ones that gave the average of five.

Well, we'll let it stand at that. A year or so later, he published a new volume of his book. In that, there's a chapter on the sealed up cards in the envelopes and they all come up to around seven. And nothing is said about the fact that for a long time they came down below five. You see, he *knows* if they come below five, he *knows* that isn't fair to the public to misrepresent this thing by including those things that prove just as much a positive result as though they came above. It's just a trick of the mind that these people do to try to spite you and of course it wouldn't be fair to publish [12].

Flying Saucers

I'm not going to talk about flying saucers very much except just this. A flying saucer is not exactly science, although some scientific people have written things about them. I was a member of General Schwartz's (?) Advisory Committee after the war, and we held some very secret meetings in Washington in which there was a thing called project SIGN. I think it's s-i-g-n. Anyway, it was hushed up. It was hardly even talked about and it was the flying saucer stuff, gathering the evidence, and weighing and evaluating the data on flying saucers. And he said, "You know, it's very serious, it really looks as though there is something there." Well, I told him afterwards—I told him this story here. I said that it seems to me from what I know about flying saucers they look like this sort of thing. Well, any-way, it ended up by two men being brought to Schenectady with a boiled down group of about twenty or thirty best cases from hundreds and hundreds that they knew all about. I didn't want them all, I said to pick out about thirty or forty of the best cases, and bring them to Schenectady, and we'll spend a couple of days going over them, and he did.

Most of them were Venus seen in the evening through a murky atmosphere. Venus can be seen in the middle of the day if you know where to look for it. Almost any clear bright day especially when Venus is at its brightest, and sometimes it's caused almost panic. It has caused traffic congestion in New York City when Venus is seen in the evening near some of the buildings around Times Square and people thought it was a comet about to collide with the earth, or somebody from Mars, or something of that sort. That was a long time ago. That was thirty or forty years ago. Venue still causes flying saucers.

Well, they only had one photograph or two photo-graphs taken by one man. It looked to me like a piece of tar paper when I first saw it and the two photographs showed the thing in entirely different shapes. I asked for more details about it. What was the weather at the time? Well, they didn't know but they'd look it up. And they got out some papers and there it was. It was taken about fifteen or twenty minutes after a violent thunderstorm out in Ohio. Well, what' s more natural than some piece of tar paper picked up by a little miniature twister and being carried a few thousand feet up into the clouds and it was coming *down*, that's all. So what could it be?" But it was going at an enormous speed." Of course the man who saw it didn't have the vaguest idea of how far away it was. That's the trouble. If you see something that's up in the sky, a light or any kind of an object, you haven't the vaguest idea of how big it is. You can guess anything you like about the speed. You ask people how big the moon is. Some say it is as big as your fist, or as big as a baseball Some say as big as a house. Well, how big is it really." You can't tell by looking at it. How can you tell how big a flying saucer is? Well, anyway, after I went through these things I didn't find a single one that made any sense at all. There was nothing consistent about them. They were all things that suffered from these facts. They were all subjective. They were all near a threshold. You don't know what the threshold is exactly in detecting the velocity of an object that you see up in the sky, where you don't know whether it's a thousand feet or ten thousand feet or a hundred thousand feet up. But they all fitted in with this general pattern, namely, that there doesn't seem to be any evidence that there is anything in them. And, anyway, these men were convinced and they ended project SIGN. And later the whole thing was de-classified and the thing was written up by the Saturday Evening Post about four or five years ago. At any rate, that seemed to be the end of it. But, of course. the newspapers wouldn't let a thing like that die. (Laughter) It keeps coming up again, and again, and again, and the old story keeps coming back again. It always has. It's probably hundreds of years old anyway.

Well, I think that's about all. If there are any questions, I'd be happy to say more.

Question Period

W.C. White: People may want to go now because it's quarter after five though I'm sure Dr. Langmuir would be glad to discuss this some more.

I was going to add another one to these characteristics. Isn't the desire for publicity another of the characteristics?

A. Well, it is in Rhine's case. There is no question about that. Rhine, I think,. *thinks* he's honest, but I know perfectly well that he—everything he says, he talks about the importance of getting more students, and the importance of having the people in his own university understand the importance of this thing and so on And then the fact that no man in his senses could discard data the way he did those things sealed up in the cards. So I don't hold a very high value on his work. Now the other people, I don't have the slightest doubt but what these men are really honest. They are sincere. They loved publicity; Allison, of course, loved to publish about new elements one after the other. These were published by the American Chemical Society; and Latimer liked to publish his little article on tritium, the first discovery of tritium. So I think that has something to do with it, but I don't think that that's the driving force. I think the driving force is quite a normal scientific desire to make discoveries and to understand things. Davis and Barnes were finding things and it was wonderful while it lasted.

Q (Liebhafsky): I just wanted to point out that perhaps the neatest comment on item four was made at the University of California when this business was discussed at the Research Conference there in about 1930 or 32. Professor Birge said that this effect was just Allison wonderland. (Laughter)

(Langmuir): Did you ever hear Latimer talk about it?

(Liebhafsky): Well, Latimer was pushing it and you've got to allow for Latimer's persuasiveness. There were people on the faculty that I'm sure never believed it.

(Langmuir): But it was funny that G.N. Lewis would believe it.

(Liebhafsky): Well, you know that there is a very close personal relationship between Latimer and Lewis.

(Langmuir): I understand that Lewis got back his ten dollars. (Laughter)

Q. How would an analysis like this apply to religious experiences?

A. Well, the method of approach to religious questions—a lot of people think you don't want to have any evidence, you want faith; and if that's your attitude why I don't think this thing applies. But if some religious performer of a certain belief tries to argue with me, my reactions would be very much like this.

Q. In setting up these criteria, you may in a way limit the possibilities of scientific investigation. It occurred to me that suppose something happened in the heavens—some astronomical event—that nobody had ever seen before. Something that happens once in a million years. Really, I mean, supposing that you could tell. It would fit the same criterion, wouldn't it?

A. No, I don't want to depend on any one of these. I've been reading the life of Pasteur. Pasteur had the idea of germs. Everybody thought that he was a fool—thought there couldn't be any sense to the subject. It took a long time before germs were believed. People believed in spontaneous generation of new forms of life. They happened spontaneously not by the introduction of spares from the outside but spontaneously—and Pasteur had to fight that. The test of time is the thing that ultimately checks this thing. In the end, something is salvaged. You can't do that while the thing is growing, while the thing is being discussed, but in the end you do know that the Allison effect is gone. It never would be anything. And that's what I mean about these other things. We've waited long enough now. This whole pattern of things fits together with the idea that you're at a threshold. You're right at the point where things are very difficult to see—that's what I want to bring out. Now, in Pasteur's experiments, when he killed anthrax in animals, he got 25 right out of 25. The sheep all died or they didn't die. There was no threshold value about it. People who didn't know anything about it might have thought so, but when they saw one experiment they were convinced.

One more question

Q. These criteria that you put down would apply very well to the theory of relativity with measurements of very small fractions of a degree of arc in the neighborhood of a bright disk of the sun.

A. Yes, well now take an example I've often thought of. There are lots of scientific instances. They go through the same sort of stage. For instance, in Laue and Bragg's theory of x-rays being electromagnetic waves. When the first reports came out you had to keep an absolutely open mind about them. You didn't know but what this was just another case of wishful thinking. But how long did it take? Within three or four years they were making precision measurements of the wavelengths of x-rays—very, very few years. Now, that's just what doesn't happen in these things. So you have to wait a little time for these things to prove themselves but I don't think that you will find that there's anything more than a superficial resemblance. Take the first experiments of the wave theory of electrons. The first evidence was very poor, and more people had to be brought in, but to me the important thing was not how it looked at the time but the quickness with which those results were resolved as contrasted to these things that hang fire and hang fire. Now the Davis-Barnes effect and the N-rays were quenched suddenly; but most of these other things go on, and on, and on, and on.

(White): I believe that this is the latest lasting colloquium we've ever had that I remember. It was a great privilege to have such a speaker. We thank you, Dr. Langmuir.

Epilogue (R.N. Hall) (presumably written in 1968)

Pathological science is by no means a thing of the past. In fact, a number of examples can be found among current literature, and it is reasonable to suppose that the incidence of this kind of "science" will increase at least linearly with the increase in scientific activity.

Professor Allison has retired, but in a recent letter he wrote that his investigations of the Allison Effect have suffered long interruptions but were never abandoned, and he spends summers and occasional weekends working on it with students at Auburn University. The effect is also being investigated under a contract with the Air Force Aero Propulsion Laboratory at the University of Dayton.(9e)

Flying Saucers are still very much with us. As Langmuir said, "Of course, the newspapers wouldn't let a thing like that die." How right he was!

References

[1] Eight months after the visit of Langmuir and Hewlett to Columbia and this exchange of letters, Barnes submitted a paper on the Davis-Barnes effect and it was published as Barnes, A.H., "The Capture of Electrons by Alpha-Particles," *Physical Review*, Vol. 35, 217 (1930).
[2] Webster, H.C., *Nature*, Vol. 126, 352 (1930).
[3] Davis, B. & Barnes, A.H., *Physical Review*, Vol. 37, 1368 (1931).
[4] Blondlot, K. *The N-Rays*. (London: Longmans, Green and Co., 1905).
[5] McKendrick, J.G., *Nature*, Vol. 72, 195 (1905).
[6] Wood, R.W., *Nature*, Vol. 70 (1904); Wood, R.W., *Physikalische Zeitschrift*, Vol. 5, 789 (1904).
[7] Seabrook, W. *Doctor Wood*. (New York: Harcourt, Brace, and Co., 1941), Chap. 17.
[8] For a review and bibliography, see Hollander and Claus, *Journal of the Optical Society of America*, Vol. 25, 270 (1935).
[9] The following references on the Allison Effect make interesting reading (a) Allison, F. & Murphy, E.S., *Journal of the American Chemical Society*, Vol. 52, 3796 (1930). (b) Allison, F. *Industrial & Engineering Chemistry*., Vol. 4, 9 (1932). (c) Cooper, S.S. & Ball, T.R.

Journal of Chemical Education, Vol. 13, 210 (1936), also pp. 278 and 326. (d) Jeppesen, M.A. & Bell, R.M., *Physical Review*, Vol. 47, 546 (1935). (e) Mildrum, H.F. & Schmidt, B.M., *Air Force Aero Propulsion Laboratory*, AFAPL-TR-66-52 (May 1966).

[10] Latimer, W.M. & Young, H.A., *Physical Review*, Vol. 44, 690 (1933).

[11] This may have referred to the paper by McGhee, J.I. & Lawrenz, M., *Journal of the American Chemical Society*, Vol. 54, 405 (1932), which contains the statement, "In December 1930 one of us (McGhee) handed out by number to Prof. Allison twelve (to him) unknowns which were tested by him and checked by two assistants 100 percent correctly in three hours." See also, Ball, T.K., *Physical Review*, Vol. 47, 548 (1935), who describe additional tests in which unknowns were identified.

[12] Some more recent discussion of Rhine's work is to be found in: (a) Price, G.R., *Science*, Vol. 122, 359 (1955), and replies on January 6, 1956. (b) Gardner, M. *Fads and Fallacies in the Name of Science.* (Mineola, NY: Dover, 1957) p.13.

28

"Science Fans" and "Amateur Scientists"
Song Tian

Song Tian (田松) is a Chinese professor of the philosophy of science. A professor at Beijing Normal University, he studied physics at Jilin University and Nanjing University before studying at the Chinese Academy of Sciences and Beijing University. He one of the foremost writers and researchers on pseudoscience in China and created the term Science Fans to talk about those amateur scientists who claim to engage in legitimate research but are actually dabbling in pseudoscience. In this essay, he discusses many of the attributes of Science Fans and their differences with amateur scientists.

What is a science fan?[1]

Tian [1] coined the term "science fans" to refer to a peculiar crowd who want to do so-called scientific research outside of the scientific community. They attempt to solve certain grand scientific questions, attempt to overthrow certain well-established scientific theories, or are devoted to establishing grand types of theoretical systems. However, they do not accept or understand the scientific community's basic paradigms of knowledge and therefore cannot have a basic communication with the scientific community. In general, their work does not possess scientific sense or value[2].

---

* Reprinted with permission of the author from Song Tian,"Science fans: A basic description and analysis of the emergence of a pseudoscience movement in China" *Natural Dialectical Research* (*Ziran Bianzhengfa Yanjiu*) Vol. 7 (2003) 56 Translated by Reginald Smith.

1 Science fans have not in general attracted the attention of the scholarly community. This paper was written using the following data 1) Mass media reports on science fans, 2) the author's own data collected on science fans, and 3) direct correspondence with science fans.

2 Regarding the difference between science fans and the scientific community the author is postulating a theory and not just engaging in a discussion.

Science fans are a very large group[3]. They are interested in almost every prominent field, especially those with ideological value. In the field of mathematics many are dedicated to proving Goldbach's Conjecture[4]. This problem attracts the largest group of science fans. Li Daqing [2], a reporter with a large and influential publication of the scientific community, *Science & Technology Daily*, calls them "Goldbach Fans." Even today, the Mathematics Division of the Chinese Academy of Sciences still receives sacks of letters and papers claiming they have proved Goldbach's Conjecture. In the field of physics, they are devoted to overthrowing the theory of relativity[5], quantum theory, or advancing new universal theories. There are even a few people researching perpetual motion machines. In the field of biology, there are those trying to advance new theories of evolution[6]. Besides this, in geology, psychology, and other scientific fields there is no lack of such people. There are some whose theories are grand beyond comparison from theories of cosmic creation to theories of *Yin* and *Yang* and the five cosmic elements (metal, wood, earth, water, and fire), from the

[3] Since there is no relevant statistical data, it is difficult to estimate the exact number of science fans. The Mathematics Division of the Chinese Academy of Sciences estimates there are several thousand Goldbach Fans.

[4] Goldbach's Conjecture is a famous unsolved problem in number theory. In 1742, the Prussian mathematician Christian Goldbach wrote legendary mathematician Leonhard Euler with the following conjecture "Every even integer greater than 2 can be written as the sum of three primes." Euler later restated it as "Every even integer greater than 4 can be expressed as the sum of two primes." Euler did this so 1 does not have to be considered a prime number.

[5] At the Chinese Academy of Sciences' History of the Natural Sciences Division, researcher Song Zhenghai organized a group called "The Strangers to Earth" and had a series of lectures. In July 2000, the First National Conference on Questions about Einsteinian Relativity was convened and published its proceedings as Second Thoughts on Relativity . Song Zhenghai was the editor (Earthquake Press, July, 2002) (in Chinese).

[6] Among the most famous proposed theories of human evolution was that of Zhu Haijun. He believes that humans walk upright because human ancestors adapted to face-to-face sexual intercourse.

food and drink of daily life to political economies. All encompassing but unable to produce concrete knowledge.

Science fans' psychological characteristics, behavioral tendencies, writing style, etc. have many common aspects. In summary, their innermost characteristic is obstinacy. They have great trust in the peculiar worth and meaning of their own "scientific conclusions." They cannot get along with the scientific community or communicate with normal people. They often live with illusions of grandeur. For example, they cannot truly understand other people's opinions and can be prone to ignore unfavorable parts of other people's opinions while exaggerating the parts they like. Sometimes they also engage in wishful thinking, for example, comparing themselves to those like Giordano Bruno[7] and Galileo Galilei. They often run into opposition and interpret this as authority exercising oppression and persecution towards an outsider.

They also have a universal characteristic of conducting an intense and vigorous search along the lines of pure idealists. They generally cannot manage a good life, and some of them even depend on their parents and their wives and children after the age of forty. However, the difficulties of life actually strengthen their conviction of sanctity and solemnity. Many of them have faith that in the end they will become a great master. This conviction gives a silver lining to their difficult lives.

Science fans popularize their "research results" largely in the following ways:

1. Writing letters to introduce their systems and complain for a redress of grievances against them (targets perhaps are academic journals, science magazines, popular media, and government officials on all levels.)
2. Giving lectures (generally university campuses are their most favored places).
3. Publishing books at their own expense

---

[7] Giordano Bruno was an Italian monk burned at the stake by the Catholic Church in 1600 for his spiritual beliefs but is falsely believed by many to have been burned for hypothesizing the existence of extraterrestrial life, a view he did hold.

4. Constructing their own research institutes[8]
5. Publishing their results on-line

Their "research theses" also have some general characteristics

1. They create many new terms, however, these terms have very little relationship to the technical terminology of the scientific community.
2. Confused logic without any direction
3. Often exaggerate the significance of their conclusions
4. Like publishing on the transcendental meaning of a question, especially in terms of national patriotism.
5. Often base their conclusion on possibilities which can only be realized in the future and cannot be achieved by modern science—especially, on the possibility of a scientific revolution that may be launched by themselves.

Besides, from the perspective of educational background, they usually have not received professional training on the field which they dedicate themselves to, nor have they undergone self-study to achieve a thorough understanding of the field. In fact, as a result of the science fans' general nature, they have become significant subjects of sociological research.

Up to now, neither scholars nor the mass media refer to them by a standard name. Some call them "citizen scientists" or "amateur scientists," both of which they like to refer to themselves as. This article's author started referring to them as "science fans." This has appeared in the media since 2000 [3]. Recently, Dr. Liu Huajie invented a new type of name, "Quack science fans" (*jianghu kexue aihaozhe*). This name is meant to inspire ridicule at their most blatant characteristics. Generally speaking, the work of science fans is not classified as pseudoscience [1] but as "science-imitating pseudoscience" or academician Hao Bolin's [4] so-called "fake science" (*yan kexue*). Yet science fans all do things their own way individually. Another usual sign of pseudoscience is that it produces serious social influence in economic or political activity of society in China. Because of these reasons, science fans are different.

[8] Feng Yiquan proposed that the Earth will leave the moon behind. He self-constructed a "philosophical geology" research department on Mount Tang (Tangshan). See South China Weekend 4/7/2000 edition. Article by Liu Xuejin "Feng Yiquan -- One man's 'theory'" (in Chinese)

Science fans' members consist of every sector of society including party cadres, teachers, civil servants, private sector employees, farmers, factory workers, and even the elites[9]. The age range is also wide with the oldest science fan being more than seventy years old. Generally though, the lower limit is around thirty years old.

2. The difference between science fans and amateur scientists

The terms science fan and amateur scientist stem from necessity of differentiation. On the surface, science fans and this other group have many similarities. Amateur scientists also are not professional scientists, also like to engage in scientific activities in their free time, and even have some obstinate people, but they accept the paradigms of the scientific community. They are also capable of engaging in progressive exchanges with the scientific community, and simultaneously are capable of creating some valuable scientific research. These people could be called true amateur scientists[10].

It can be difficult to differentiate between the literal meaning of the terms "science fan" and "amateur scientist" in Chinese. Here one must not prepare to regard it as a language question, but accept the names as they are. Amateur scientists are also a very large group, addressing those engaged in a wide scope of scientific activity. For example, there are many amateur astronomers who spend their free time to observe astronomical phenomena. There are many amateur astronomers who are able to discover new heavenly bodies. However, on the other hand, they do not intend

[9]The highest level science fan probably is the Chinese People's Political Consultative Committee (a legislative advisory committee in the Chinese government composed of both Communist Party members and non-members) committee member Huang Wei. He has continuously claimed to have designed a perpetual motion machine. See Fascinating Limits--The Recollections of a War Criminal Administration Bureau Chief 1998: People's Liberation Army Press (in Chinese)

[10] Two types of translations of "science hobbyists" have been discussed on the Rainbow Bridge Science Education Forum. A lot of friends on web provided many helpful suggestions and information. The author is grateful for their assistance.

to overthrow the theories of modern astronomy, nor do they expect to establish a new overarching cosmic theoretical system[11].

Amateur scientists at the same time have not become a special crowd, but are still common members of our society. The basic differences between the two groups consist of the following: amateur scientists and the scientific community conduct regular collaborations. Among amateur scientists, these collaborations occur very frequently. This is similar to the relationship between amateur and professional chess players. But science fans not only cannot conduct a progressive exchange with the scientific community, they also cannot carry out a mutual exchange among themselves. "Goldbach fans" are examples, each one of them independently creates their own "theoretical system" and wishes that other people would read about and understand their system. However, none of them want to read or even understand the "theoretical systems" created by any others. Therefore although there are many people, they are not a collective but rather an assemblage.

References

[1] Tian, Song, "Scientism, Anti-Science, and Pseudoscience" *Natural Dialectical Research* Vol. 9, 19 (2000). (In Chinese)
[2] Li, Daqing, "Goldbach Fans 'Proofs' get a Cold Shoulder from Experts. Goldbach's Theory is Still Not Laid to Rest" *Science and Technology Daily* March 29, 2002. (In Chinese)
[3] Tian, Song, "I see Science Fans" *Science Times* March 27, 2000 B3. (In Chinese)
[4] Han, Bolin, "Pseudoscience and Fake Science", *Science (Shanghai)*, Vol. 2 (2002).

[11] On February 1, 2002 the International Astronomical Union's Commission on Minor Planets and Comets gave the name "Ikeya-Zhang" to the comet which was independently discovered by two amateur astronomers, one Chinese (Zhang Daqing) and one Japanese (Kaoru Ikeya). Zhang Daqing is an ordinary laborer in Henan province. He found this comet using a self-constructed 20cm reflector telescope.

29

The Art of Scientific Investigation: Excerpt on Scientific Ethics

William Ian Beardmore Beveridge

There are certain ethical considerations which are generally recognised among scientists. One of the most important is that, in reporting an investigation, the author is under an obligation to give due credit to previous work which he has drawn upon and to anyone who has assisted materially in the investigation. This elementary unwritten rule is not always followed as scrupulously as it should be and offenders ought to realise that increased credit in the eyes of the less informed readers is more than offset by the opprobrium accorded them by the few who know and whose opinion really matters. A common minor infringement that one hears is someone quoting another's ideas in conversation as though they were his own.

A serious scientific sin is to steal someone's ideas or preliminary results given in the course of conversation and to work on them and report them without obtaining permission to do so. This is rightly regarded as little better than common thieving and I have heard a repeated offender referred to as a "scientific bandit." He who transgresses in this way is not likely to be trusted again.

Another improper practice which unfortunately is not as rare as one might expect, is for a director of research to annex most of the credit for work which he has only supervised by publishing it under joint authorship with his own name first. The author whose

---

* Reprinted from W.I.B. Beveridge, The Art of Scientific Investigation (New York: W.W. Norton, 1957) 145-148 under Creative Commons Attribution-ShareAlike License 3.0

name is placed first is referred to as the senior author, but senior in this phrase means the person who was responsible for most of the work, and not he who is senior by virtue of the post he holds. Most directors are more interested in encouraging their junior workers than in getting credit themselves. I do not wish to infer that in cases where the superior officer has played a real part in the work he should withhold his name altogether, as over-conscientious and generous people sometimes do, but often it is best to put it after that of the younger scientist so that the latter will not be overlooked as merely one of "and collaborators." The inclusion of the name of a well known scientist who has helped in the work is often useful as a guarantee of the quality of the work when the junior author has not yet established a reputation for himself It is the duty of every scientist to give generously whatever advice and ideas he can and usually formal acknowledgment should not be demanded for such help.

Some colleagues and myself have found that sometimes what we have thought to be a new idea turns out not to be original at all when we refer to notes which we ourselves made on the subject some time previously. Incomplete remembering of this type occasionally results in the quite unintentional annexing of another person's idea. An idea given by someone else in conversation may subsequently be recalled without its origin being remembered and thus be thought to be one's own.

Complete honesty is of course imperative in scientific work. As Cramer said,

> *In the long run it pays the scientist to be honest, not only by not making false statements, but by giving full expression to facts that are opposed to his views. Moral slovenliness is visited with far severer penalties in the scientific than in the business world.* [1]

It is useless presenting one's evidence in the most favourable light, for the hard facts are sure to be revealed later by other investigators. The experimenter has the best idea of the possible errors in his work. He should report sincerely what he has done and, when necessary, indicate where mistakes may have arisen.

If an author finds out he cannot later substantiate some results he has reported he should publish a correction to save others either being misled or put to the trouble of repeating the work themselves, only to learn that a mistake has been made.

When a new field of work is opened up by a scientist, some people consider it courteous not to rush in to it, but to leave the field to the originator for a while so that he may have an opportunity of reaping the first fruits. Personally I do not see any need to hold back once the first paper has been published.

Hardly any discovery is possible without making use of a knowledge gained by others. The vast store of scientific knowledge which is to-day available could never have been built up if scientists did not pool their contributions. The publication of experimental results and observations so that they are available to others and open to criticism is one of the fundamental principles on which modem science is based. Secrecy is contrary to the best interests and spirit of science. It prevents the individual contributing to further progress; it usually means that he or his employer is trying to exploit for their own gain some advance made by building on the knowledge which others have freely given. Much research is carried out in secret in industry and in government war departments. This seems to be inevitable in the world as it is today, but it is nevertheless wrong in principle. Ideally, freedom to publish, provided only that the work has sufficient merit, should be a basic right of all research workers. It is said that occasionally, even in agricultural research, results may be suppressed because they are embarrassing to government authorities [2]. This would seem to be a dangerous and shortsighted policy.

Personal secrecy in laboratories not subject to any restrictions is not infrequently shown by workers who are afraid that someone else will steal their preliminary results and bring them to fruition and publish before they themselves are able to do so. This form of temporary secrecy can hardly be regarded as a breach of scientific ethics but, although understandable, it is not commendable, for free interchange of information and ideas helps hasten the advance of science. Nevertheless information given in confidence must be respected as such and not handed on to others. A travelling scientist visiting various laboratories may himself be perfectly honourable in not taking advantage of unpublished information he is given, but may inadvertently hand on such information to a less scrupulous individual. The traveller can best avoid this risk by asking not to be told anything that is wished to be kept confidential, for it is difficult to remember what is for restricted distribution and what not.

Even in the scientific world, unfortunately, one occasionally encounters national jealousies. These are manifest by lack of appreciation or acknowledgment of work done in other countries.

Not only is this to be deplored as a quite indefensible breach of ethics and of the international spirit of science, but it rebounds on the offenders, often to the detriment of themselves and their country. The person failing to appreciate advances in science made elsewhere may be left in the backwater he deserves, and he shows himself a second-rate scientist. Among the great majority of scientists there exists an international freemasonry that is one of the main reasons for faith in the future of mankind, and it is depressing to see this marred by petty selfishness on the part of a few individuals.

References

[1] Cramer, F. *The Method of Darwin. A Study in Scientific Method.* (Chicago: McClurg & Co., 1896).

[2] Hughes, D.L. "The present-day organisation of veterinary research in Great Britain: Its Strength and Weaknesses." *Veterinary Record*, No. 60, 461 (1948).

INDEX

Come learn more about the Citizen Scientists League!

http://citizenscientistsleague.com

Also check out the upcoming book from member George Hrabovsky co-authored with renowned physicist Leonard Susskind, The Theoretical Minimum: What You Need to Know to Start Doing Physics, to be published in January 2013 by Basic Books.

About the Editor

Reginald Smith has been active in the amateur science community for over ten years with both the Society for Amateur Scientists and the Citizen Scientists League. His research focus over the years has been mainly in physics including plasma physics, complex networks, and the physics of complex systems such as the Internet, finance, and biology. He has published over 12 peer reviewed scientific papers and wrote extensively on amateur science topics. He studied at the University of Virginia receiving a B.S. in Commerce and double majoring with a B.A. in physics. He also received his Masters in Business Administration from the Sloan School of Management at the Massachusetts Institute of Technology.

NOTES

NOTES

NOTES

NOTES

NOTES

NOTES

NOTES

Made in the USA
Lexington, KY
06 August 2012